KU-257-089

WITHDRAWN

638.
109
747
DAY

LIVERPOOL
JOHN MOORES
UNIVERSITY

Library Services

Accession No	1886029	
Supplier PQ		Invoice Date
Class No		
Site A	Fund Code LBS1	

LIVERPOOL JMU LIBRARY

3 1111 01518 7113

HONEYBEE HOTEL

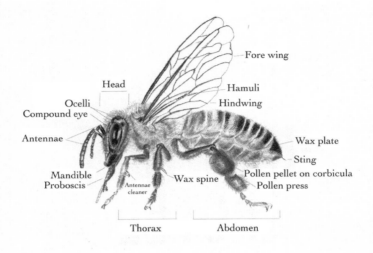

Fore wing

Head

Ocelli

Compound eye

Antennae

Hamuli

Hindwing

Wax plate

Sting

Mandible

Proboscis

Antennae cleaner

Wax spine

Pollen pellet on corbicula

Pollen press

Thorax

Abdomen

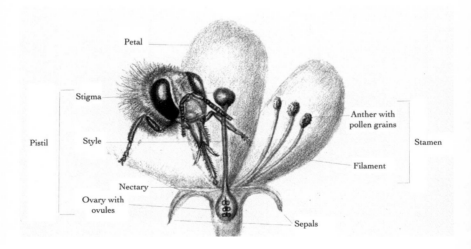

Petal

Stigma

Pistil

Style

Anther with pollen grains

Stamen

Filament

Nectary

Ovary with ovules

Sepals

HONEYBEE HOTEL

THE WALDORF ASTORIA'S ROOFTOP
GARDEN *and the* HEART *of* NYC

LESLIE DAY

JOHNS HOPKINS UNIVERSITY PRESS
BALTIMORE

© 2018 Johns Hopkins University Press
All rights reserved. Published 2018
Printed in the United States of America on acid-free paper
2 4 6 8 9 7 5 3 1

Johns Hopkins University Press
2715 North Charles Street
Baltimore, Maryland 21218-4363
www.press.jhu.edu

Library of Congress Cataloging-in-Publication Data

Names: Day, Leslie, 1945– author.
Title: Honeybee hotel : the Waldorf Astoria's rooftop garden and the heart of NYC /
Leslie Day.
Description: Baltimore : Johns Hopkins University Press, 2018. | Includes
bibliographical references and index.
Identifiers: LCCN 2017056665| ISBN 9781421426242 (hardcover : alk. paper) |
ISBN 1421426242 (hardcover : alk. paper) | ISBN 9781421426259 (electronic)
| ISBN 1421426250 (electronic)
Subjects: LCSH: Bee culture—New York (N.Y.) | Roof gardening—New York
(N.Y.)—History. | Waldorf-Astoria Hotel (New York, N.Y.)—History. |
Honey plants.
Classification: LCC SF523.3.D39 2018 | DDC 638/.1097471—dc23
LC record available at https://lccn.loc.gov/2017056665

A catalog record for this book is available from the British Library.

Frontispiece: Honeybee anatomy and flower anatomy by Leslie Day

Special discounts are available for bulk purchases of this book.
For more information, please contact Special Sales at
410-516-6936 or specialsales@press.jhu.edu.

Johns Hopkins University Press uses environmentally friendly book materials,
including recycled text paper that is composed of at least
30 percent post-consumer waste, whenever possible.

Book design by Kimberly Glyder

FOR THE CHEFS:
DAVID GARCELON, PETER BETZ, AND CALOGERO ROMANO

The possession of knowledge does not kill the sense of
wonder and mystery. There is always more mystery.
—Anaïs Nin

CONTENTS

PREFACE

The idea for this book came through a family connection. My nephew, Peter Betz, at that time the executive sous chef at the Waldorf Astoria Hotel, told me that his boss, the newly hired director of culinary, David Garcelon, wanted to put beehives on the roof of the hotel. I was teaching science to middle school students at the time. Peter knew that my students studied pollinators and flowers, so the following year he arranged to give my husband and me a tour of the garden and apiary. The stories Peter told of the dedication of the staff at the Waldorf to the hives were spellbinding. Their completely volunteer efforts to build the hives and the raised garden beds were an inspiration. As a researcher, writer, and educator of the natural history of New York City, I felt a powerful connection to these bees and to the people at the hotel. Here were hardworking men and women, many of them immigrants, who cared about these tiny insects. And then there were the bees, whose lives were completely devoted to their community.

Peter told me that the hotel would share the honey and the produce grown on the roof with the homeless shelter and food pantry across the street at St. Bartholomew's Episcopal Church. The lives of the workers at New York's Waldorf Astoria Hotel, the laborious bees, and the connection to the charitable work of St. Bart's helping New Yorkers in desperate need of shelter and food—this was a story that needed to be told.

The lives of humans and bees are intertwined and often parallel. Nothing on our planet escapes this construction. Except for the first three chapters on the history of the Waldorf Astoria Hotel, the structure of this book follows a pattern that alternates between our species, *Homo sapiens*, and the industrious honeybee, *Apis mellifera*.

During the course of writing this story of the Waldorf Astoria Hotel and its honeybee garden, the Hilton Corporation sold the Waldorf, its flagship hotel, which Conrad Hilton had called "the greatest of them all," to China's Anbang Insurance Company. Readers will learn more about the fate of New York City's most historic and iconic hotel in the epilogue.

Entomologists will forgive me, I hope, in my spelling of *honeybee* as one word. They prefer two, as *Apis mellifera* is but one of many bee species, though it is special in many regards. I have adopted the single-word spelling because it is a common literary usage and often preferred among beekeepers. If this were a technical volume, which it most assuredly is not, I would have followed their preference, and used *honey bee*. Let me also add that I have been as careful as possible to get the facts right. Mistakes, if there are any, are mine alone. The entomologists and beekeepers I mention in this book were very helpful through my interviews with them and their published research and books. If I misconstrued anything, forgive me.

HONEYBEE
HOTEL

WELCOME TO THE WALDORF

> The opening of the new Waldorf Astoria is an event in the advancement of hotels, even in New York City. It carries great tradition in national hospitality . . . marks the measure of [our] nation's growth in power, in comfort and in artistry . . . an exhibition of courage and confidence to the whole nation.
> —President Herbert Hoover, broadcasting from the White House, on opening day of the Waldorf Astoria, October 1, 1931

Trudging through six inches of slushy snow and carefully walking around black ice, I slogged east on 49th Street from where the D train deposited me at Rockefeller Center. I was on my way to meet David Garcelon, director of culinary at New York City's famed Waldorf Astoria Hotel. I wanted to talk about his brainchild: a rooftop honeybee garden. As I walked through the winter urban landscape, I thought of the bees waiting out the winter snugly in their hives and felt the irony of choosing a day in late January to meet David, when the temperature was 14 degrees Fahrenheit.

The idea for this book began during a breakfast with my nephew, Peter Betz, then executive sous chef, who had been working his way up the chef's ladder at the Waldorf for twenty years. Eventually, Peter became executive chef of the hotel.

On a morning several years before my January trudge, my husband and I had met up with Peter; it was a decidedly warm day in the autumn of 2011. Thoughts of it crossed my mind as the cold wind of January bit into me at a street corner. We met Peter, his wife, Moriko, my brother-in-law Eizo, and his wife, Ruth, for an incredible breakfast buffet at Oscar's Brasserie, the historic Waldorf Astoria restaurant. After feasting on smoked salmon, bagels, waffles, bacon, eggs, and fruit, Peter took us on a tour of the Waldorf Astoria. He guided us through the massive kitchens, the four-story Grand Ballroom, and all the famous public event spaces where presidents, heads of state, writers, artists, and musicians have performed, spoken, and inspired the world. His enthusiasm was contagious, and he shared wonderful stories about each historic room. Then he took us to the twentieth floor and out to the roof, where he talked about David Garcelon's plans to establish honeybee hives and a full garden for the bees and chefs.

The summer of 2013 Peter again took us up to the roof. There we found a fully flowering chefs' garden and six honeybee hives that sheltered as many as 300,000 bees producing honey and pollinating the flowering plants. Looking south we could see the magnificent art deco Chrysler Building with its glittering, stainless steel crown of terraced and radiating arches. Looking north below us was the magnificent tiled Byzantine dome of St. Bartholomew's Church. Thirteen beds of herbs, spices, vegetables, and their edible nasturtium, lavender, thyme, dill, cilantro, daylily, squash blossom, chives, and basil flowers, surrounded by flowering apple and cherry trees,

Goldsmiths and Silversmiths Company of London's Clock, 1893 Chicago World Exposition. *Waldorf Astoria Archives*

created a fragrant oasis twenty stories above Park and Lexington Avenues. The bees were covered in pollen from the flowers and were busy sopping up as much nectar from each blossom as their tiny honey stomachs could hold. As a naturalist and a New Yorker, I was enthralled by the art deco beauty of this most historic of hotels, and of the natural wonders that thrived on its roof twenty stories above the bustling city.

✄ ▰▰▰ ✄

Thoughts returning to the present, I stomped my boots and made my way inside the Waldorf's entrance. I could see David waiting for me by the World's Fair Clock in the Waldorf's main lobby. It's a common meeting place, an icon that has brought countless people together for more than one hundred years. I looked from David to the clock. It is an astounding artistic timepiece; it forces you to pause. Its age, size, details, and chimes make most visitors to the Waldorf gape as they take it in.

Located in the exact center of the main lobby, the clock stands nine feet tall, is ornately carved bronze, and sits atop an octagonal marble and mahogany base. It was created by the Goldsmiths and Silversmiths Company of London and commissioned by Queen Victoria as a gift to America for the 1893 Chicago World's Fair. It weighs two tons and has the bas-relief carved faces of Queen Victoria, Benjamin Franklin, George Washington, Abraham Lincoln, Andrew Jackson, Ulysses S. Grant, Grover Cleveland, and William Henry Harrison. It's topped with a shiny bronze Statue of Liberty.

A description of the clock, written by the Committee on Awards of the World's 1893 Columbian Commission, is worth reading for its timeless quality:

> Goldsmiths and Silversmiths Company, London: Among the many grand productions of this company was an "Exposition clock" specially designed and manufactured for the World's fair in 1893. The case was eight feet high and octagonal with elegantly embossed and richly gilded ornaments, the cotton plant and flower being the principal subjects. The case had eight panels, representing the sports swimming, running, yachting, cycling, baseball, trotting, and jumping, with a view of Brooklyn Bridge. Each panel was surmounted by a portrait of Washington, Lincoln, Grant, Franklin, Jackson, Harrison, and Cleveland and a medallion portrait of Queen Victoria. The clock had four dials show-

ing English, American, French and Spanish time. About four feet from the floor on a platform around the clock, were twelve figures representing players in cricket, rowing, shooting, polo racing, lacrosse, boxing, running, tennis, football and wrestling, which revolved in procession as the clock striked each quarter hour. Above, at the sides of the dials, four columns supported brackets with vases, between each of which were figures signifying progress in art, science, industry, and engineering. The American eagle was shown above each dial. At each hour the American and English anthems were played, the time being denoted by a chime of 8 bells, the Westminster chimes on four gongs, and the hour on a gong. All can be repeated at will, imparting to the whole work a realism and effectiveness they claim has never before been attained in any similar production. The clock movement and other mechanism showed the good work characteristic of English manufacture.

When the Chicago World's Fair ended, John Jacob Astor IV bought the clock and put it inside the lounge off the 34th Street horse carriage entrance of his original Astoria Hotel. In 1902, a small bronze Statue of Liberty, a gift from France to Astor, honoring the hospitality the people of France had received at his hotel since 1897, was placed atop the clock, where it stands today.

David Garcelon, wearing his white chef's outfit and tall chef's hat, was standing by the clock. We shook hands, and he led me into the Peacock Alley Restaurant so that we could sit in comfort for the interview. He was greeted by the maître d' and waitstaff with big smiles and affection. I could tell that he was one of those bosses that didn't cause apprehension in workers but, instead, evoked respect and fondness—the kind of boss we all hope to have at least once or twice along the way.

The Peacock Alley Restaurant is an extension of the Waldorf's plush lobby, and it is lovely: filled with art deco treasures, ceilings etched with glass panels, patterned parquet floors, and murals of white peacocks. The walls are covered with French walnut burl veneer panels inlaid with ebony. The pillars are Moroccan marble, and the ceiling and cornices are covered in gold and silver leaf. Everywhere you look your eyes fall on beauty, and history.

Even the restaurant's name has meaning. Peacock Alley was the name given to a 300-foot-long corridor with amber marble and mirrored walls that

The original Peacock Alley. *Waldorf Astoria Archives*

connected two hotel buildings owned by John Jacob Astor's great-grand-sons, the feuding Astor cousins, William Waldorf Astor and John Jacob Astor IV. It was a place to see and be seen. On weekends, up to 36,000 men and women walked the alley, admiring themselves and one another in the mirrors, prompting the name "Peacock Alley" as a description of the strut-ting to and fro. The Peacock Alley Restaurant was named for that famous corridor.

David is a tall, dark, lean, and handsome man in his fifties. Like me, he is a swimmer. Unlike me, he is on his feet for ten to twelve hours a day, over-seeing a staff of 140 chefs who prepare meals for hotel guests in the 1,413 rooms, several restaurants, ballrooms, and event spaces within this huge building. He is the calm at the center of the storm, coming across as gentle and generous, never making you feel rushed even though you know he has a high-pressured schedule.

He spent his childhood exploring the natural world of his hometown, St. Stephens, New Brunswick, Canada. It is a small, rural community that shares the St. Croix River as a border with Calais, Maine. The St. Croix flows south and southeast between the two towns, eventually meeting Passa-maquoddy Bay in the Bay of Fundy. Twice each day the water rises and falls an amazing fifty feet at this spot, creating the most dramatic tides anywhere in the world.

"I spent time outdoors, summer, fall, winter, and spring," David said, responding to my first question. "My mother and grandmother cooked from scratch from their gardens. We ate with the seasons: fiddlehead ferns and asparagus in the spring; corn, blueberries, and strawberries in summer; fresh corn in the fall; and seafood throughout the year."

Eating from the garden and from the nearby rivers, lakes, and sea instilled in him a love of nature and fresh food. But David didn't cook much at home. His first serious introduction to the kitchen came from a summer job as a high school freshman. He was fourteen and worked at a neighborhood pizzeria.

"We made everything from scratch, and it would get very busy," David explained. "The shift would fly by, with getting your *mise en place* ready (a term used by professional chefs, meaning arranging all of your ingredients beforehand) and then flying through service trying to keep up. I got addicted to that adrenaline rush of a busy service and then the satisfaction of having done it well and survived."

He continued to work part time at the pizzeria after summer ended and through that cold Canadian winter. Then he went on to get jobs in other restaurants throughout his high school years. By the time he was sixteen, he had decided he wanted to be a chef.

"To my astonishment, my parents were supportive," he told me. "I had expected them to push me toward university. Instead, my father bought me a copy of *Larousse Gastronomique* for Christmas in 1983. I read it like a novel."

David had three key ingredients that formed his foundation: a supportive family, a strong urge to work hard, and a desire to give through food. He would build a career in nurturing others through food that would one day lead him to the "greatest of them all"—the Waldorf Astoria Hotel in the heart of New York City.

To understand the foundation of this celebrated and historic hotel, we have to travel back in time more than 250 years to Waldorf, Germany, and learn about the son of a butcher whose name was Johann Jakob Astor.

THE FIRST AMERICAN TYCOON

JOHN JACOB ASTOR IN NEW YORK CITY

> Could I begin life again, knowing what I now know,
> and had money to invest, I would buy every foot of land
> on the Island of Manhattan.
> —John Jacob Astor, 1834

The history of the Waldorf Astoria Hotel is the story of some of the city's most significant, creative, and complex figures, starting with John Jacob Astor. Born Johann Jakob Astor in 1763, in the small town of Waldorf, Germany, he was the fourth and youngest son of Johann Jacob Astor, a butcher. Wanting something more for his life than working in his father's trade, he left Waldorf in 1779 at the age of sixteen, as revolution raged in the American colonies. The first destination of his life's amazing journey was London, where his brother George had established himself as a successful musical instrument craftsman, building pianofortes, clarinets, and flutes.

Astor was young, poor, and in a strange land. But his sister-in-law Elizabeth, George's wife, taught him to speak, read, and write English. He changed the spelling of his name from Johann Jakob to John Jacob Astor as a way of fitting in. Though he had left school at age thirteen, he was a whiz at math and kept his brother's books. By age seventeen, he had learned quite a bit about how to run a business. In 1783, after the American Revolution had ended, Astor was on the move again. This time he joined another brother, Henry, in New York. By 1784, New York was not only the center of commerce and trade for the colonies but had been made the new nation's capital. Even after Washington, DC, became the capital, New York remained the trading center of the United States, and people were flocking to the city from Europe.

Astor's brother Henry, like his father, was a butcher. He helped John Jacob find housing and work with a baker friend. His first job was delivering Dieterich's freshly baked German bread. Astor became involved in the German Reformed Church of New York and was quickly named the congregation's treasurer, managing all of their finances. Delivering baked goods lasted only a short time. Astor used his savings to import musical instruments from London made by his brother George. Astor shrewdly guessed that after several years of war—with nothing imported from London during the Revolution—musical instruments would be in demand.

In an effort to make more money, he began to investigate the fur trade, which was becoming the most lucrative business in the city.

A series of cold winters in Europe caused a run on fur hats and coats by Europeans, almost resulting in the extinction of the European beaver. When Henry Hudson returned to Europe with large quantities of beaver pelts, the Dutch West India Company jumped at the chance to exploit the market. They sent trading ships to New York filled with goods for the Native Americans in exchange for beaver pelts. The North American species of beaver had made the fortunes of many, and Astor wanted in. He got his start in the fur business at the bottom, by scraping and cleaning pelts for John Bowne in Lower Manhattan. Eventually, he was promoted to keep the books, and, as he had done in London, he learned everything he could about his new trade.

In 1785, Astor married Sarah Todd, a young Scottish American whose family had been in New York for generations. Sarah had close ties to family and friends who were involved in the shipping trade with Europe, and she and her husband set up a fur shop in their house. Sarah also introduced her husband to ship captains who helped him establish an international trade in furs and musical instruments. This was the beginning of Astor's trade empire. Astor began traveling across North America, buying and selling furs, while Sarah ran the shop and business at home. In 1808, he and Sarah created the American Fur Company and soon had practically monopolized the fur trade. Their company became the most successful business in the young United States. Their fur trading empire spread through the Midwest and into the Western Frontier through trading posts. Astor named one in Oregon, Fort Astoria (it's now the picturesque port town of Astoria, in Northwest Oregon).

America was expanding westward, and the young nation was growing economically and geographically. At the same time, destructive forces were being unleashed on Native Americans and the American beaver by fur traders. Alcohol, disease, depletion of game, and warfare between tribes over hunting grounds decimated the Native American population. The American beaver was almost hunted out of existence in territory after territory.

In 1830, Astor took his fortune from trading furs and invested it in real estate. He bought huge tracts of land, including farms in what is now Midtown Manhattan, from 5th Avenue to the Hudson River. His hunch that the city would expand from Lower Manhattan, as more and more people immigrated to the city, paid off. The population of New York City was 25,000 when Astor arrived in 1783. When he died in 1848, it was more than a half-million people, all needing places to live and work. "The City's Landlord" was one of the nicknames given to Astor. After his death, his Midtown real estate holdings were developed into Times Square, the Theater District, and the Garment District.

In 1836, John Jacob Astor opened his first hotel. It was named the Park House, but everyone called it the Astor House. Built in the Greek Revival style, it sat on Broadway between Vesey and Barclay Streets, across the street from City Hall Park. At six stories and with 300 rooms, maid service, and seventeen bathrooms (an incredible luxury for that period), it was a spectacle with running water pumped to the upper floors by a unique steam engine and lit with gas.

It was the talk of the town and attracted wealthy families, politicians, and celebrities, among them Davy Crockett, Ralph Waldo Emerson, Henry Wadsworth Longfellow, Charles Dickens, and Daniel Webster. William James was born in the hotel. In 1855, Emerson, a great supporter of Whitman's *Leaves of Grass*, an ode to New York City and one of the first great urban epics, invited Whitman to the Astor House, where he was staying. In 2010, 155 years later, Waldorf Astoria mixologist, Frank Caiafa, would create a cocktail in honor of the great poet, naming it Leaves of Grass. About the hotel, Walt Whitman wrote, "Among the elder buildings, only the Astor House, in its massive and simple elegance, stands as yet unsurpassed as a specimen of exquisite design and perfect proportion." In 1861 President-elect Abraham Lincoln stayed at the hotel on the way to his inauguration, and the night before, standing on the hotel portico, he gave a speech to thousands of supporters.

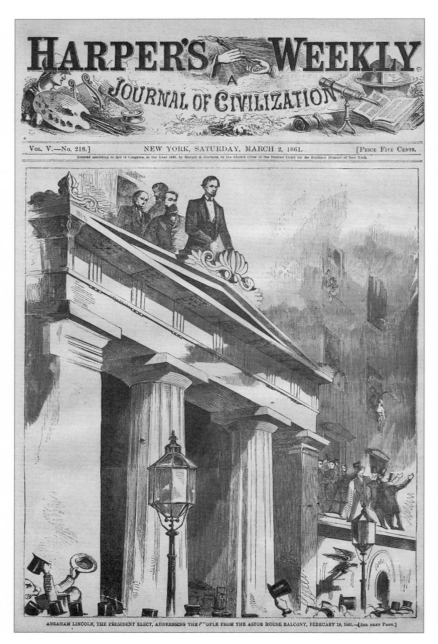

Abraham Lincoln speaking to supporters from the Astor House, March 2, 1861.
Photo provided courtesy of The Philadelphia Print Shop, Ltd.

In 1864, the Astor was one of several hotels set on fire by backers of the Confederacy. After he was assassinated in April 1865, Lincoln's funeral procession moved past the Astor Hotel to City Hall, where his body lay in state.

Astor was always fighting the perception that he was a heartless, money-grubbing tycoon. To combat this image, he donated some of his money to a variety of causes. He financially supported artists and writers, such as John James Audubon, Washington Irving, and Edgar Allan Poe. He gave money to the elderly, women's and children's shelters, hospitals, and schools. He bought and supported the famous Park Theater, one of the major theaters in America at that time. When it burned down in 1820, Astor immediately rebuilt it as a gift to the city.

With guidance from his friend Joseph Cogswell, the prominent American educator and librarian, Astor left $400,000 in his will to create the Astor Library. After Astor's death, Cogswell traveled throughout Europe, collecting tens of thousands of books and meeting the great writers, scientists, and artists, including Goethe, Alexander von Humboldt, Lord Byron, and Walter Scott. The library opened its doors in 1854 with a holding of 80,000 volumes to be used for research by the public, which made it the first public library in America. It was not a lending library; researchers had to stay inside the building. It was built on Lafayette Street, which was renamed Astor Place. In 1895, thirty-seven years after Astor's death, the Astor Library was joined with the Lenox Library and the Tilden Library to form the New York Public Library. If you stand on 5th Avenue and gaze up at the façade of the left side of the building, you will see engraved there: "The Astor Library, founded by John Jacob Astor for the Advancement of Useful Knowledge MDCCCXLVIII."

When John Jacob Astor died in 1848 at the age of eighty-five he was worth $20 million, making him the richest man in America, and the country's first millionaire.

3

BEFORE THE EMPIRE STATE BUILDING, THE WALDORF ASTORIA STOOD HERE

> During the era that was invariably referred to as the
> Gay Nineties, when the nation was still expanding and its people
> still expansive, Peacock Alley set the pace. A now forgotten journalist
> of the time once wrote that in that fabulous corridor every day
> was Easter Sunday and every moment was a promenade.
> — Karl Schriftgiesser, *Oscar of the Waldorf*

John Jacob Astor's son William Backhouse Astor Sr. (1792–1875), his grandsons, and great-grandsons all grew Astor's original earnings to create an empire worth hundreds of millions of dollars. They built hotels and thousands of stores and apartment houses in the growing city. By 1890, his two famous great-grandsons, forty-one-year-old William Waldorf Astor and twenty-five-year-old John Jacob Astor IV, lived next door to each other on the northwest corner of 33rd Street and 5th Avenue and the southwest corner of 34th Street and 5th Avenue, respectively.

According to William Alan Morrison, author of *Waldorf Astoria*, there was a "walled private garden and a turgid sea of mutual dislike" between their mansions. To put it plainly, they couldn't stand each other. William believed he was the head of the family and thought of his younger cousin as a loafer. It tormented him that Caroline Schermerhorn Astor, the mother of the younger Astor, was the crowned queen of New York society and spoken of as *the* Mrs. Astor; she was the woman who created the powerful list of the "four hundred" elite families during the Gilded Age. William believed that

Peacock Alley Society, Fin de Siècle. *Waldorf Astoria Archives*

his wife, Mary Paul Astor, should have that title. But high society sided with "Lina" Schermerhorn Astor as the "official Mrs. Astor," resulting in William finally moving his family to England. Before he left he demolished his house and built the thirteen-story Waldorf Hotel in its place, forever commercializing that corner of 5th Avenue and bringing the loud life of the city to what had been a quiet, residential block.

The hotel was designed by architect Henry Janeway Hardenbergh, the architect of two famous New York City landmark buildings: the Dakota Apartments on Central Park West and 72nd Street and the Plaza Hotel on 5th Avenue and 59th Street.

The Waldorf Hotel, at first scorned by its residential neighbors for ruining a good neighborhood, was managed by George Boldt whose motto was "The guest is always right." Boldt had been the proprietor of Philadelphia's fashionable Bellevue-Stratford Hotel and was hand chosen by William Waldorf Astor to manage his new hotel. On the Waldorf's opening day, Boldt held a benefit for St. Mary's Hospital for Children, a favorite charity of New York society, knowing full well that the society ladies, led by Caroline Schermerhorn Astor, had to show up. The new "uptown" hotel became a huge success, and four years later, Caroline's son, John Jacob Astor IV, with the help of George Boldt, convinced his mother to move uptown and replace their mansion with the seventeen-story Astoria Hotel.

Caroline Schermerhorn Astor's residence next door to the Waldorf Hotel, 1894.
Library of Congress

At first, the hotels were completely separate buildings, but at Boldt's suggestion they were connected with a mirrored and gilded alley, later dubbed "Peacock Alley" by the press because of the strutting thousands decked out in their finery. The 1,000-room Waldorf-Astoria was born, the embodiment of luxury at the *fin de siècle*. From 5th Avenue, New Yorkers gazed at its balconies, turrets, chimneys, and gorgeous red-tiled gables. Its interior was filled with treasures collected from around the world.

Journalist Robert Stewart described his arrival at the hotel in 1899 in *When the Astors Owned New York* this way:

> Think of it! You arrived tired, dusty, irritable. Your bag is whisked out of your hand, and you are conducted through a brilliant hall . . . Presto! You find yourself in a bijou of a suite, your trunks awaiting you, with a bed which simply beseeches you to lie on it, and with a porcelain tiled bathroom all your own. You press

The Waldorf Hotel, 1893.
Waldorf Astoria Archives

The Astoria Hotel, 1897.
Waldorf Astoria Archives

one button in the hall; electric lights flash up. You press another; a maid or valet knocks to unpack your luggage and help you to dress. You press a third; a hall boy appears, like the slave of Aladdin's lamp, to execute any possible command monsieur may issue from fetching a glass of iced water to ordering a banquet served up to you.

The food was the talk of the world. Several renowned restaurants within the hotel were run by Oscar Tschirky, called "Oscar of the Waldorf." He was a famous maître 'd'hôtel hired by Boldt to run the hotel's kitchens from its opening in 1893. Oscar published *The Cook Book by "Oscar" of the Waldorf* in 1896, 900 pages of his chefs' recipes, including many we enjoy today, such as the Waldorf salad, eggs Benedict, red velvet cake, and Thousand Island dressing, the latter named for New York State's Thousand Islands, where George Boldt built his "castle." Oscar served at the Waldorf-Astoria for half a century: both the original hotel and then the Park Avenue icon, until his retirement in 1943.

Boldt became a millionaire himself and not only built his famous castle, now open to the public, but owned thousands of acres of land in the Thousand Islands. His Wellesley Island Farms, a sprawling and productive farm on an island in Alexandria Bay near the St. Lawrence River, had dairy cows,

THE NEW YORK HERALD.

THE TITANIC SINKS WITH 1,800 ON BOARD; ONLY 675, MOSTLY WOMEN AND CHILDREN, SAVED

MOST APPALLING DISASTER IN MARINE HISTORY OCCURS WHEN WORLD'S LARGEST STEAMSHIP STRIKES GIGANTIC ICEBERG AT NIGHT

John Jacob Astor IV goes down with the Titanic. *Smithsonian Library Digital Photo Archive*

beef cattle, hogs, horses, sheep, hens, chickens, ducks, turkeys, and geese. Boldt's model farm supplied his hotels—the Waldorf-Astoria in New York and the Bellevue-Stratford in Philadelphia. It delivered enough produce, beef, chicken, pork, lamb, duck, turkey, goose, milk, bread, and eggs to feed thousands of guests every day.

Not everything surrounding the early Waldorf-Astoria was joyful. The best-known tragedy affiliated with the hotel was the fate of John Jacob Astor IV. He had been in a loveless marriage for many years, but after his mother, *the* Mrs. Astor, died in 1908 in her mansion on 65th Street and 5th Avenue (now the site of Temple Emanu-El), he felt free to divorce his wife in 1909. He then married seventeen-year-old Madeleine Talmage Force in 1911. Their relationship was the talk of society. Many were against divorce, and Episcopalian ministers in the city refused to marry them, so they married in Newport, Rhode Island.

Senate investigation at the Waldorf-Astoria into the sinking of the Titanic, April 19, 1912. *Waldorf Astoria Archives*

Eating lunch on the scaffolding of the new Waldorf-Astoria Hotel. *Waldorf Astoria Archives*

They honeymooned in Egypt and Europe and decided to return home to New York on the maiden voyage of the *Titanic*, departing from Cherbourg, France, on April 12, 1912. At almost midnight, April 15, the *Titanic* was moving at 21.5 knots when it hit an iceberg. As the ship began to sink, the captain called for the lifeboats to be filled, women and children first. After getting his pregnant bride onto a lifeboat, Astor stayed behind. According to some reports, he had refused a seat in the lifeboat that contained his wife. He was one of the 1,503 who died on that historic day.

Looking south on Park Avenue at the Waldorf-Astoria Hotel, 1929. *Waldorf Astoria Archives*

Days later, the very first US Senate investigation of the sinking of the *Titanic* took place at the Waldorf-Astoria Hotel.

By the 1920s, wealthy New Yorkers started moving uptown, and the Waldorf followed them. The Astor family sold the 5th Avenue hotel's site to real estate developers in 1929, and it was torn down to make way for the most symbolic and iconic of all New York City skyscrapers: the Empire State Building. Lucius Boomer, then president of the Waldorf-Astoria Hotel Corporation, retained the rights to the name, and together with T. Coleman du Pont, built and managed the new Waldorf-Astoria.

From 1929 to 1931, the hotel rose from 301 Park Avenue, taking up an entire square block covering East 49th Street, Park Avenue, East 50th Street, and Lexington Avenue.

From its opening day in 1931 until 1963, it was the tallest hotel in the world. It was designed by architects Schultze and Weaver, who also designed the Pierre Hotel and Sherry Netherland Hotel, and its art deco beauty ultimately gave the building's exterior landmark status by the New York City Landmarks Preservation Commission. It was designed as a city within a city. And it is, still today, recognized as one of the world's most gorgeous and

——— COLE PORTER ———

WALDORF TOWERS RESIDENT, 1934-1964
NOTED LYRICIST AND PLAYWRIGHT

WHILE IN RESIDENCE, HE PENNED MANY OF
HIS FAMOUS PLAYS AND PIECES,
INCLUDING "DON'T FENCE ME IN" AND
"YOU'RE THE TOP, YOU'RE A WALDORF SALAD."

HIS STEINWAY GRAND PIANO CAN BE
FOUND ON THE COCKTAIL TERRACE
IN OUR PARK AVENUE LOBBY.

Cole Porter plaque outside his apartment in the Waldorf Towers. *Leslie Day*

historic hotels: a magnet for politicians, world leaders, celebrities, and international tourists.

Oscar had a new restaurant within it named for him: Oscar's Brasserie. The new hotel had 1,413 hotel rooms. The Waldorf Towers, soaring from the twenty-eighth to the forty-second floor, had 181 rooms with 115 of them suites leased by and named for their famous tenants. One was the Cole Porter Suite, which in 1964, the year Porter died, became the home of Frank Sinatra.

The Royal Suite was named after the Duke and Duchess of Windsor, who lived there from 1941 to 1961 whenever they were in New York. The MacArthur Suite and the Churchill Suite were named for their famous inhabitants. The most expensive room, the Presidential Suite, was designed with the Colonial-style furniture meant to evoke the White House. It was the residence of President Herbert Hoover for thirty years following his retirement and was used by every American president, up to Barack Obama, as their headquarters during the United Nations General Assembly.

The retractable roof of the Starlight Room and its outdoor terraces filled with trees let guests dine beneath the stars. Decades later the twentieth-story roof would be made into a chefs' garden populated by honeybees.

Starlight Roof, 1935. *Waldorf Astoria Archives*

LIFE OF THE HONEYBEE

FROM EGG TO ADULT

> The bee is more honored than other animals, not because
> she labors, but because she labors for others.
> —Saint John Chrysostom

The admiration and respect humans feel for the industrious *Apis mellifera* stems from thousands of years of observing them and depending on their work to pollinate the flowering plants that produce so much of our food. Every part of the honeybee's body is designed for its work to collect and process a flower's resources. In turn, many flowers use the opportunity of the bee's visit to pass pollen to another of its species in one of nature's most gentle acts of procreation. The entire surface of a honeybee, including its compound eyes (which contain a hair where each facet meets), can become covered with pollen. Every visit to a flower is sure to haul away thousands of pollen grains on their hairs, bristles, barbs, and spikes. They fly to flowers to collect pollen and watery nectar, which they turn into honey—a process of evaporation that takes place inside the hive—to feed their young, their drones, their queen, and themselves. The purpose of their lives is to serve their community.

Within every honeybee society are three types, or castes, of bees: (1) the queen, who lays all of the eggs; (2) the female workers who do all of the labor; and (3) the male drones who mate with neighboring queens.

The queen, who can lay between 1,000 and 2,000 eggs each and every day, inspects the empty comb cells before she lays one egg, and one egg only, in each cell. She is a discerning mother and does not abide messiness. She is looking for an exquisitely sculpted beeswax cell, spotlessly cleaned by her daughters, in which to deposit each one of her progeny.

The worker bees that build the comb determine the number of cells to construct for workers, drones, and queens-in-waiting based on the needs of the colony. Small cells will house fertilized eggs that will become the workers. Larger cells are for unfertilized eggs that will become the male drones. The largest cells are for the fertilized eggs that will become queens-in-waiting.

Each tiny, white, translucent egg is laid standing on one end deep within its waxy cell. Over the next two to three weeks, the egg will metamorphose into a larva, then a pupa, and, finally, an adult bee. When a worker bee's transformation ends, the young adult honeybee emerges and immediately starts her first job, thoroughly cleaning her tiny room of debris so that the queen will use it to deposit an egg.

Each egg is slightly curved, with the wider end becoming the bee's head and the slender end developing into its abdomen. The queen has inserted enough yolk into every egg so that the embryo is nourished before it emerges and is fed by its sisters. Within hours, as it develops and grows heavier, the tiny egg slowly sinks and lies curled on the floor of its cell. It takes three days for the membranous covering of the egg to dissolve, revealing the baby bee, now called a *larva*. In Latin, the word *larva* means "ghost." And at this stage, the baby bee is ghostlike, glistening and white, lying curled up on the bottom of her cell.

Larval bodies are marvelously equipped for one thing and one thing only: to feed. They have no legs, eyes, antennae, wings, or stinger. They have only their mouths, which they use to lap up the enormous amount of food given to them by nurse bees so that they can feed and grow. The rest of their body is made up of salivary glands, midgut and hindgut, and excretory organs so they can digest, poop, and develop. They don't stay small for long. Their sisters feed them pollen and honey almost 150 times a day. Their tiny homes, the brood cells, are cleaned over and over each day as the nurse bees tend to the babies. A nurse bee may visit a cell 1,500 times a day to remove waste and food debris and to feed the growing child within.

The amount and quality of the food fed to the larva determines whether the bee will develop into a worker or a queen. When they first hatch, all three castes—workers, queens-in-waiting, and drones—are fed royal jelly brood food, consisting of secretions from the glands located in the heads of nurse bees: hypopharyngeal (just below the pharynx) and mandibular glands. After a few days, workers and drones are fed simply honey and pollen: honey for carbohydrates in the form of sugars and pollen for protein, lipids, vitamins, and minerals, while queens-in-waiting are only fed royal jelly. The larvae of each caste are fed differently, according to their needs, but they are always fed the products of delicate and beautiful flowers: pollen and honey, plus secretions from the nurse bees' brood food glands. Together, these provide the growing larvae with everything they need in order to go through metamorphosis and function as highly productive adults.

Larvae produce pheromones (chemicals released into the air that influence the behavior of others) letting the nurse bees know they are hungry. These pheromones stimulate the hypopharyngeal brood food glands of the nurse bees to perform their early-in-life tasks, the feeding and caring for larvae.

Larvae shed their skin, or exoskeleton, five times: once each day throughout the larval stage. Growing rapidly, they shed their external skin like a too-tight shirt that a child has outgrown. After five molts, they metamorphose into their pupal stage. On the fifth day, the attending worker bees seal the larvae inside their cells with a cap made of beeswax.

It takes several days for the larvae to spin their cocoons, during which time they are not fed. They fully straighten their bodies with their heads toward the cell's waxen cap and start to weave cocoons using their spinnerets. They produce silk directly from glands that become salivary glands when they attain adulthood. Their white bodies darken during this stage of life, and they develop a head, eyes, antennae, mouthparts, a thorax, legs, an abdomen, and wings. Their eye color changes from white, to pink, to purple, to brown, and finally to black. The thorax turns yellow. Their internal organs and muscles go through enormous changes. This is the last stage before the final molt.

Once they have shed their final larval skin, it takes up to twenty-four hours for their outer layer, or cuticle, to harden, for their hairs to stiffen, and for their stinger to function. While this is taking place, they use their mandibles to chew holes in the wax covering of their cell. As they perforate the wax, they remove tiny pieces and attach them to the rims of their cell. This wax will be collected by janitor bees and reused to cap other cells. Nothing is wasted.

Once out of their cells, they rest as their hairs dry and they stretch out their wings and antennae. From egg, to larva, to pupa, to adult, the journey takes sixteen days for queens, twenty-one days for workers, and twenty-four days for drones. During this period, the queens-in-waiting grow 1,700 times the weight of their eggs, the workers 900 times, and the drones 2,300 times.

New worker bees are fed by the nurse bees for a short time and then forage on their own for pollen left in empty cells. New drone and queen bees cannot feed themselves immediately after they emerge from their cells because they are not fully mature: their reproductive organs are still developing. They beg for food from the nurse bees by thrusting their tongues forward toward the mouth of a passing bee, who will open her mandibles and regurgitate a drop of fluid from her honey stomach, which the hungry bee laps up.

The life of the worker bee is rich in the skills she develops from the time she is four or five days old until she dies. Her many jobs help her tens of thousands of sisters, her hundreds of brothers, and her queen survive.

A GOOD CHEF IS LIKE A MUSICIAN

DAVID GARCELON'S JOURNEY

> As someone who has worked in hotels for the majority of
> my career, the opportunity to be part of the Waldorf Astoria is
> something I have always dreamed of. Leading kitchens where
> numerous iconic dishes were created instills a powerful drive
> to expand upon and innovate the hotel's culinary offerings.
> — David Garcelon

The Culinary Institute of Canada opened its doors in 1983 on Prince
Edward Island, known to locals as the Garden of the Gulf for its
gorgeous pastoral setting and productive agricultural land. It is also
known as the setting for a beloved series of books by Lucy Maud Montgom-
ery, starting with *Anne of Green Gables*. It's the story of the life of eleven-year-
old orphan Anne Shirley as she grows up on Prince Edward Island, on the
Cuthbert family farm in the 1890s.

David applied and was accepted to the Culinary Institute of Canada.
He started in the fall of 1984 as part of its second entering class. The chef
instructors were from Germany, England, Austria, and France, and the
students were taught classic French cuisine. "It was a school ahead of its
time in that part of the world, and, in my opinion, it is still one of the best
culinary schools in the world. I was the youngest student at eighteen but had
more experience in a real kitchen than almost anyone else."

It was a life-changing experience for the young chef. When he completed
the courses, he had hopes of returning home to be near family and friends

and applied to the Algonquin Resort in St. Andrews by-the-Sea, New Brunswick, Canada. Built in the 1880s, this towering Victorian-era complex overlooks Passamaquoddy Bay.

> The Algonquin Hotel wouldn't even interview me. I ended up, to my dismay, at the Windsor Arms Hotel in Toronto. So I moved to the big city by myself, and was lonely but working in an amazing European-style kitchen that was producing some of the best food in the city. I loved the work but missed my friends and hated the big city so I took an offer to work with one of my old instructors from culinary school who was then in Halifax. Several of my friends from school were there. I worked there for eighteen months, and then three of us decided to move to Vancouver, which was getting ready for the 1986 World Expo. It was really hard for me to leave home, but it was the best decision I ever made for my career. It was there that Chef Wolfgang Leske gave me my first sous chef position. He was an incredibly hardworking and organized chef who influenced me greatly.
>
> I remember talking to my grandmother, Nan, when I was agonizing over this move to Vancouver. She told me, "David, go live your life."

This simple advice helped him decide to move away from his roots and set out on a professional career.

Although growing up in a rural town influenced David's love of nature and garden foods, he described the ways that the positions he held in western Canada influenced him even more:

> It was working in Canadian hotels in pristine national parks that instilled in me a reverence for the land. I think more than anything this respect was due to my experience working for Canadian Pacific hotels—now the Fairmont hotels—for twenty years. They were pioneers in the environmental movement. They were the first to put cards in the room, saying, "If you don't want your towels washed everyday just hang them up" and "Tell us if you don't need your sheets changed every day." They were the first hotel chain to use recycling bins. This was very much the culture

of these spectacular resorts in the national parks of the Canadian Rockies.

David ran the kitchens in some of the most beautiful hotels in that part of Canada, which are among the most spectacular in the world. "They very much saw themselves as stewards of the environment," he said. "I was lucky enough to live and work in the Rockies. My kids grew up in Jasper Park Lodge. It really was a privilege to live in a national park. The people who work there have a sense of stewardship for the park."

Jasper National Park in the Canadian Rocky Mountains of Alberta is the home to diverse wildlife. Bald eagles, black and grizzly bears, elk, bighorn sheep, harlequin ducks, moose, and wolves live in the park year-round. The lakes are pristine. The park covers 4,200 square miles and is one of North America's largest nature preserves. It is a United Nations Educational, Scientific and Cultural Organization (UNESCO) World Heritage site. UNESCO was founded to recognize and protect the world's most important cultural and natural areas. Only 197 sites in the world have been chosen for their natural significance to humanity, and Jasper is one of them. To be chosen for this title and protection a site must meet the following criteria:

- contains superlative natural phenomena or areas of exceptional natural beauty and aesthetic importance;
- is an outstanding example representing major stages of Earth's history, including the record of life, significant on-going geological processes in the development of landforms, or significant geomorphic or physiographic features;
- is an outstanding example representing significant on-going ecological and biological processes in the evolution and development of terrestrial, fresh water, coastal and marine ecosystems, and communities of plants and animals; and
- contains the most important and significant natural habitats for in-situ conservation of biological diversity, including those containing threatened species of outstanding universal value from the point of view of science or conservation. (http://whc.unesco.org/en/criteria/)

When David worked there, and in other magnificent national park hotels, he was strongly influenced by the beauty of these natural areas and by the need to take care of them, even while running a business within them.

The Jasper Park Lodge has been in business since the early 1920s and has drawn countless visitors, including notables such as Bing Crosby, Marilyn Monroe, Joe DiMaggio, Sir Arthur Conan Doyle, King George VI and Queen Elizabeth, the Queen Mother, and Queen Elizabeth II and Prince Philip. One day, not very far in the future, David would be director of culinary at a famous New York City hotel named the Waldorf Astoria, where these same people had been guests. He was on his way.

David talked about the beauty he found in cooking:

> Creativity is an interesting concept when it comes to chefs. I think many people believe it is something you either have or you don't have. Really what happens is that we spend years learning how to cook—some happens in school but most of it happens on the job learning motor skills, techniques, flavors and the experience of cooking with hundreds of different ingredients, using different methods and hopefully picking up insights from those we work with. Gradually you are able to reach into the depth of experience that you have and create dishes and develop your own style. I think it is really more like a musician learning an instrument over many years before he or she is able to truly "create" something new.

Over the course of his career, David has been inspired by several chefs.

> When I was a young chef in the 1990s, and first had my own kitchens, Charlie Trotter was a big influence. Chef Trotter was one of the first celebrity chefs out of Chicago. He was concerned with sourcing and knowing where ingredients came from. Are they local? Are they sustainable? Today, these questions are mainstream, but in the 1990s, I started incorporating these culinary values into my kitchen in the Canadian Rockies. Food writer and culinary activist Anita Stewart was another huge inspiration. When I first met Anita, in 1995, I was experimenting with local dishes and ingredients, which became known as Rocky Mountain

cuisine. She was a great source of encouragement and knowledge. She is revered in the Canadian food scene.

The desire to use local and sustainable foods led David to reach out. "We started working with local farmers," he explained, "and with an agricultural college that had a horticultural program."

Olds College is a one-hundred-year-old agriculture school in Olds, Alberta, a productive agricultural area on the Canadian prairie. Vast fields of golden wheat and buttery-yellow canola crops are grown on the Canadian prairies. Cattle number in the millions. It is also the beekeeping capital of Canada. Canola flowers depend on honeybees for pollination. Alberta boasts about 300,000 honeybee colonies that feast on the flowers of canola and other plants and produce most of Canada's honey.

David got to know a professor at the college and invited him and his students to the hotel kitchen. David walked them through the "fridges," showing them what kinds of food the hotel used.

As David led the tour of agricultural students, they expressed amazement at what they saw: fresh basil from Israel and Hawaii and tiny baby zucchini from South America. They asked, "Why would you want such little zucchinis? When we grow vegetables, we want them big." David then showed them the hotel's plates, how they arrange the baby zucchinis and why it's important to have things a certain size. "They saw what the hotel was buying from other places in the world," David said, "and they knew that this produce, like the zucchini, and the beautiful, edible blue borage flowers, which are popular right now, could be grown in their greenhouses."

This was an aha moment for David, the professor, and the students. They realized the college could grow vegetables for the hotel, and David saw the benefits of a direct connection between the producer and the chef/customer, without the middlemen. David then looked at me intently and said, "We could really talk to each other. That's powerful, I think."

6

LIFE OF THE FEMALE HONEYBEE

JANITOR, FORAGER, AND EVERYTHING IN-BETWEEN

> How doth the little busy bee
> Improve each shining hour,
> And gather honey all the day
> From every opening flower!
> How skilfully she builds her cell!
> How neat she spreads the wax!
> And labors hard to store it well
> With the sweet food she makes.
> In works of labor or of skill,
> I would be busy too . . .
> —Isaac Watts

After they emerge from their cells, the worker bees begin their division of labor approach to the jobs that must be done to keep the hive humming. In their very short lives, measured in weeks if they are spring or summer bees, or months, if they are autumn bees, they will start to work when they are only hours out of their cells. Skills that take us years to learn, young bees pick up almost instantly. They learn to be janitors, sanitation workers, baby nurses, chefs, food processors, packagers, construction workers, heaters, air conditioners, guards, and foragers.

These working girls are not multitaskers. To the contrary, they master each job by working with laser-like focus day in and day out on that job— and that job only. When their bodies mature, they move on to the next vital task, changing roles as they age, and master their new skill within hours. If their work is completed, they rest, often for long periods of time. Honeybee scientist Mark Winston notes in his book *Bee Time*, "Bees in unstressed colonies are restaholics rather than workaholics, but when required they can ramp up by compressing the normal time frame of work into a shorter, more intense period."

Newly metamorphosed worker bees begin fulfilling their duties within a few hours after emerging from their cells. The first jobs they perform are janitors and sanitation workers, cleaning out cells by removing waste and cocoons, old pollen, the wax caps of cells, and dead larvae and adults.

Several days later they become baby nurses and chefs when they are able to produce food in two specialized glands in their heads: the hypopharyngeal and mandibular glands. This specially prepared food, called royal jelly, is for their larval sisters and brothers, often delivered mouth to mouth. For the first couple of days of larval life, all larvae—workers, drones, and queens-in-waiting—are fed royal jelly. This brood food is made up of a clear liquid derived from the nurse bees' hypopharyngeal glands, mixed with honey, digestive enzymes, and water, plus a milky-white component secreted by the mandibular glands. The hypopharyngeal glands are located behind the face and are connected to the base of the tongue by a duct and produce the proteins, lipids, and vitamins added to the brood food for developing larvae. As workers switch jobs from nurse bees, the hypopharyngeal glands shrink or are resorbed. If they once again engage in caring for larvae, the glands will enlarge. By the third day, they feed the worker larvae mostly liquid from their hypopharyngeal glands. During the fourth and fifth days, the larvae are fed pollen and honey directly. By day five, they are receiving heavy doses of pollen, rich in protein that will help build all of their developing cells, tissues, and organs.

Between the ages of ten and fifteen days, their next job is that of food pantry workers when they take nectar from returning foraging bees and store it in cells.

They become construction workers at about sixteen days after metamorphosis, when they secrete wax from the specialized wax glands in their abdomens and use the wax to build hexagonal cells, each beautifully sculpted to exact specifications. In these cells, either the brood will be raised or nectar and pollen will be stored. Worker bees produce all the beeswax used to construct combs. Pairs of wax glands are found below the bee's fourth, fifth, sixth, and seventh abdominal segments, hidden beneath overlapping, smooth, platelike structures called wax mirrors. These wax glands grow when the bee is secreting wax and shrink when the bee no longer performs this task. Initially, the wax is liquid and collects on the wax mirrors where it hardens into scales that look like translucent sugar glass. These waxy building blocks are then pulled up by the hairy brush on the bee's hind legs and transferred to her mandibles (jaws). She then uses her mandibles and forelegs to place the wax where it is needed.

At three weeks they become air conditioner bees and will help cool and ventilate the colony at the hive entrance by fanning their wings. And at three weeks, the workers become defenders of the colony, guarding the hive entrance against invaders. Guard duty means that their sting glands are producing large amounts of alarm pheromones and their mandibular glands, once used to help produce brood food, are now also producing alarm pheromones.

Several glands located in various parts of the bee's body are responsible for sending signals, or pheromones, out to the rest of the colony. The Nasonov gland is the bee's scent gland and is found beneath the tergite (top part) of a bee's last abdominal segment. Strategically placed at the entrance to the hive, workers lift their abdomens in the air, expose this gland, and release its "come hither" pheromone, which carries a citrusy scent, to help guide incoming foraging bees and young adult workers on their orientation flights into the nest entrance. Workers also release Nasonov pheromones to guide other foraging bees to food or water sources. During swarming, Nasonov pheromones enable scout bees to guide their colony to a new nest site.

The mandibular glands are large, paired glands on either side of the worker bee's head, attached to its mandible by ducts and extending to the base of their antennae. Secretions from these glands in young workers help produce royal jelly, but as they age out of their nurse bee roles, the mandibular glands produce alarm chemicals that smell, to some, like blue cheese. The mandibular gland is the first line of defense. When guard bees sense an intruder nearing the hive, they release their alarm scent, which signals bees in and near the hive that a trespasser is moving toward them. This is one reason why it's best to move slowly around a beehive. Quick movements can trigger activation of alarm pheromones.

Sting glands are the second stage of the alarm pheromone system. If bees need to, they will sting the intruder. Once they do, their barbed stinger (technically called a sting) is torn from the abdomen and the bee dies. When the stinger and its poison sac are inside the invader, the stinger releases strong alarm pheromones, which smell like bananas and act as a bull's-eye to guide other bees to the intruder. Pulled to attack by the pheromones, sisters of the first attacker continue to sting the intruder until it is dead or driven away from the hive.

In *The Sting of the Wild*, entomologist Justin O. Schmidt writes that stings have evolved from the egg-laying organ known as the ovipositor in insects. Only females have stings and only the worker bees die when they sting because their stingers are barbed and are caught inside their victims. The queen has a smooth sting that does not get stuck inside her victim; therefore, she can use her stinger to kill other would-be queens. She then deposits hundreds of thousands of eggs each year from the base of her sting.

Schmidt has created a Pain Scale for Stinging Insects, with values from 1 to 4 (1 is the least painful and 4 the most painful). He rates the pain from a honeybee sting as a 2—moderately painful. Here is his description:

> Western honey bee (*Apis mellifera*): Burning, corrosive, but you can handle it. A flaming match head lands on your arm and is quenched first with lye and then sulfuric acid.

Changes in the work of glands correspond to changes in jobs. The needs of the hive dictate the development and resorption of glands in the worker bees. One study demonstrated that mandibular, hypopharyngeal, and wax glands enlarge in young workers during their first two weeks as adults, corresponding to when they are feeding brood and constructing comb.

The workers' final job in life is foraging. Foragers spend their days flying from flower to flower and returning to the hive with supplies of nectar and pollen. When they become collectors of nectar and pollen, their foraging glands enlarge during the next two weeks of their lives. As the salivary glands expand and the hypopharyngeal glands start to produce a particular enzyme, the worker is able to process nectar.

While they are away from the hive, each worker must be on constant lookout for predators. The world is full of them: birds, predatory insects, and spiders. Using pheromones, landmarks, and the position of the sun, the foragers are able to find their way home even after traveling as much as six miles from their hives. For a small insect, that is quite a distance.

We have mouths for drinking and muscles inside our throat and esophagus for swallowing. Honeybees have a proboscis (pro-bah-sis), a jointed, tonguelike organ that can be neatly folded inside the mouth or extended to mop up nectar inside a flower. It is made up of a narrow tube within a wider tube. The narrow tube is used to suck up small amounts of floral nectar. The wider tube is used to suck up larger amounts of water or honey, such as when they need to fill their honey stomachs in preparation for leaving the nest when they are about to swarm or when they detect smoke. The smaller tube used for collecting tiny amounts of liquid found inside flowers is equipped with a hairy spoonlike tip called the *glossa* (Greek for tongue) that mops up small drops of nectar. The glossa has taste receptors so the bee can taste the sweet nectar it is drinking. The proboscis is protected by a hard covering.

The proboscis is also used for communication among bees. They can use it to lick pheromones from their queen and then share this chemical information with their sisters through *trophallaxis*, a process in which food and pheromones are exchanged mouth to mouth between bees. This informs the workers in a colony that their queen is alive and well.

Along with the proboscis, other mouthparts are their jaws, or mandibles. They are strong and spoon shaped to allow the bee to perform such vital jobs as eating pollen, cutting and shaping wax for comb, and cutting and shaping propolis, a sticky plant resin they use to seal and caulk openings in the hive. Their mandibles are used to feed larvae and the queen, to groom themselves and one another, to fight intruders, and to grab debris and dead bees and drag them out of the hive. These miraculous little insects are equipped with some great tools.

AN IDEA BLOSSOMS

CHEF GARCELON'S FIRST HOTEL GARDEN AND APIARY

> Whenever I take people onto a roof you can always tell
> who the gardeners are. Most people go right to the edge and
> look out at the view. But gardeners are not looking at the
> view at all; they are looking at the garden.
> — David Garcelon

After his years in the Canadian Rockies, David spent some time at the Fairmont Southampton Hotel in Bermuda before returning to Canada and meeting his first hotel rooftop garden. "When I got back to Toronto," he said, "I was still with Fairmont hotels and their flagship hotel is the Royal York. They had a garden there for many years. It was on the fourteenth floor on the roof in downtown Toronto surrounded by glass and steel. When I got there I knew that I needed to do something more with it. They got a bit of PR, but they weren't really using the garden. It was planted, but not well, and I knew if we were going to use it and do it well, I needed some advice."

Chef Garcelon is good at seeking advice. He knew that many people had a lifetime of experience and success beyond his circle of colleagues, and he started asking around for someone who knew about plants. "I was really lucky to find Marjorie Mason, a master gardener and retired public school teacher. Her family had a small greenhouse north of Toronto, and she had a popular radio show teaching people about gardening." He immediately reached out to her.

Marjorie was great—an old-style schoolteacher. She helped us plant with the seasons so the garden was always producing. And she made it look good aesthetically. She grew up on a farm in Ontario and had that in her blood. She gardened as a hobby her entire life. She owned her own greenhouse and taught people there. She also took them on tours to the Chelsea flower show and worked with us and helped us make the garden beautiful. There had always been interest in the Royal York's chefs' garden from a PR point of view, but once it became beautiful, there was tremendous interest to see it. Most people are interested in seeing a garden on the roof of any tall building because it's cool to be there. We brought a lot of people up there, and it really took off. We had a lot of publicity.

When I asked him how the idea to put beehives in that garden came to him, David remembers the moment and place.

It would have been the summer of 2007. We were up in the garden. I had a strong apprenticeship program, and I was up there weeding with the apprentice chefs. It struck me, that sunny day, that there were all these ladybugs and butterflies and bees in the garden. I just remarked that it was incredible that they could find the garden. Downtown Toronto is all glass and steel. There's no Central Park or any garden for miles around. How do they know? How do they know to look here? And there were monarchs and all kinds of butterflies and tons of honeybees. I thought, what if we had our own beehive? I knew about beehives. When I was a kid we had neighbors in town with beehives and I knew keeping bees wasn't a scary thing. We found an organization, the Toronto Beekeepers Cooperative. It was fate. I contacted them after an internet search.

Then I had coffee with Cathy Cosmo who was their president. The cooperative is a group of people in Toronto interested in keeping bees. But at that time, with construction all over the city, they faced being kicked out of their sites. So they had to relocate their bees out to a university—an hour and a half away from the city. They couldn't look after the bees and people in the city couldn't learn about the bees, so it was really fortuitous timing that we reached out to them. They were very interested in the roof garden space and said that they didn't know if this had ever been done before, but they thought it could work. The well-being of the bees was very important to them, and they weren't sure that having the hives fourteen stories off the ground would work, but they were interested in trying. The cost wasn't that much. If you build the hives it's not that expensive. The beekeepers' cooperative had the expertise to teach us. I had twelve apprentices I could put to this task. There are beehive kits where you get all the parts and build them. It's a bit of work, but it's fun. It was an investment of a few thousand dollars. We built six hives.

David emailed the chief engineer of the hotel, who responded immediately with a very long email about all the reasons they shouldn't put honeybees on the roof. "He talked about how there'd be bees in the rooms," recalled David, "and bees in the ventilation system and it would be dangerous. There were insurance considerations and I was thinking, 'Oh my God.'"

The hotel's general manager emailed David in response and wrote succinctly: "Please respond to the chief engineer's concerns about them getting into the rooms and ventilation system." The general manager said he'd follow up on insurance. He said that if those two concerns were satisfied, he'd approve it.

David talked to several beekeepers and to Mylee Nordin, the Toronto Beekeepers Cooperative's staff beekeeper, who said that although people worry about bees coming inside, the only reason a bee would come into a building is if there were flowers in there. "The bees aren't going to go after a can of soda pop left by an open window," David explained.

They are not like hornets or other wasps. A queen wouldn't want to make her nest inside a building. Honeybees get a bad rap because they are often confused with hornets. The insurance company said that it was okay as long as we had signs with instructions that if somebody got stung they should call security. The general manager approved the hives, and it was a big deal when we brought in the bees. People were worried about the hives becoming public knowledge, and we were very secretive about it. We didn't tell anyone we were doing it. The Royal York Hotel, like the Waldorf Astoria, is prominent, and anything that happens there is scrutinized. We built the hives over the winter. They were up on the roof. Very few people even knew they were there, and one day in the spring the bees arrived.

I remember meeting the beekeepers at the back door of the hotel. No one, not even the GM, knew that that was the day we were getting the bees. I was there waiting with my apprentices to unload the truck when it arrived with several hives and, unbelievably, two out of the three national newspapers were there. To this day, I don't know how they knew. It scared us to death, but we got some great press. It's kind of a cool story. You're wheeling a flatbed cart with twenty or thirty thousand bees down the hallway, past rooms where people are sleeping, and I remember being really worried at the time, "Oh, God, what if one falls over?" We put the bees in their hives and immediately a couple of things happened. We got a lot of press in the city, all of it positive, and we got a lot of interest.

David started to take guests on tours, which ended up increasing business in the hotel. They started to invite guests to afternoon tea in the garden. "We told them that on weekends at two and four o'clock the chef would take them up to the roof garden for tea," David told me, "and that increased business. The other thing that happened was the Toronto Beekeepers Cooperative's phone started to ring off the hook; people wanted to join. They had to start a waiting list, so they expanded and added another site. It was a win, win, win. It was great. We did a honey beer there as well. We were the first hotel in the world, as far as I know, that had bees on the roof. It was a really great experience. And then I came here."

8

LIFE OF THE MALE HONEYBEE

THE DRONE

> The drone bee dies soon after the wedding night.
> —Georgian Proverb

lthough the queen fertilizes close to 95 percent of her eggs with the sperm she has stored from the drones she mated with on her nuptial flights, the other 5 percent are laid as unfertilized eggs and become her sons, the drones. While the queen and every worker have thirty-two chromosomes: sixteen from the queen and sixteen from one of the males with which she mated, her sons have no father and thus only half the chromosomes.

Drone cells are covered with a dome-shaped cap, giving the cell more room to house these larger bees. Worker bees build them next to one another, which makes it easier for the queen to deposit unfertilized eggs, one after another. The drone larvae are fed more food than worker larvae and the food contains a greater diversity of proteins than the workers get, which helps develop the larger bodies and flight muscles drones will need for their own mating flights.

Once males metamorphose into young adults, they hang around the center of the nest where it is warmest and where there is the largest congregation of nurse bees to attend to their needs. After a few days, they move to the edge of the nest, where the honey cells are located, and begin feeding directly from the honey stores. There can be several hundred drones in a colony during the warm days of the spring and summer season when they are actively seeking out queens from neighboring hives to mate with.

The drones' large wings and massive muscles allow them to fly as fast as twenty-two miles per hour when they chase a queen. Drones have enormous compound eyes, the better to see, approach, and mate with a queen during flight. In addition, they have three small eyes, the ocelli, that face forward (rather than upward like their sisters), which allow them to better see the queen in front of them. They also have large antennae with thousands more olfactory receptors than their sisters' antennae; these also help them to detect queens. Drones can detect a queen's pheromone from as far as 200 feet away.

Drones never mate with a queen from their own colony. They are built to find queens from neighboring colonies and then die during the act of mating. It's a fast and furious life, which often ends unfulfilled. The unsuccessful drones are those who never mate. In autumn, those still in the nest when the weather turns cold, the failed Romeos, are herded to the nest entrance and physically forced out by their sisters, the workers. Any of the drones looking to reenter the hive are brutally rejected. Without any purpose to the community, they are no longer needed. Aimlessly wandering about, they eventually succumb to hunger, thirst, and cold. With winter approaching there is a limited supply of honey to ensure survival of the colony, certainly not enough to feed extra mouths.

The successful drones mature in time to take flight on a sunny, windless day. At the ripe age of twelve days, they leave home in search of the nearest congregation area, a place where drones meet queens. Why drones meet where they do is still a scientific mystery, but drones find these sites, which are often used year after year. Each male joins other drones, sometimes thousands of them, as they wait for the arrival of the queens.

Smelling the air with their antennae and looking for her with their eyes, drones eventually detect a queen by sensing her pheromones and her long, golden or black body. They set off, each drone trying to zoom past the others. The first to reach her is usually about fifty feet off the ground. He grabs her with his forelegs and climbs up her body while she flies, embracing her with all six legs. He then inserts his endophallus, the bee equivalent of a penis (the analogy ends here), into her open sting chamber. The erection of his endophallus induces paralysis of the drone and his body flips backward with the force of his ejaculation. The tip of his endophallus remains inside the queen, breaking off from his body. He falls to the ground and dies. The queen keeps flying with the drone's endophallus still inside her. It is believed that the dismembered body part helps stop the sperm from leaking out. The next drone to mount her must work a bit harder. He will remove the dead lover's gift and insert his own endophallus into her open sting chamber. Some fifteen to twenty-five drones will share their genes with a single queen. When she has enough sperm to fertilize up to a million eggs or more, she ends her participation in the ritual, flying home and leaving twenty dead lovers scattered across the land.

If a drone does not mate, he will fly home to his hive, rest, feed, and then try again and again until he mates with a queen, or dies trying.

TWENTY STORIES HIGH

A VISION OF THE WALDORF'S HONEYBEE GARDEN

> The greatest gift of the garden is the restoration of the five senses.
> —Hanna Rion

Gardens are healing places. The public gardens of New York City attract millions of visitors each year. Every borough has at least one major public garden: the New York Botanical Garden and Wave Hill in the Bronx, the Brooklyn Botanic Garden, and the Queens Botanical Garden. In Manhattan, there are the Conservatory Gardens in Central Park, the High Line Gardens, and the Heather Garden in northern Manhattan's Fort Tryon Park. Staten Island is home to the Snug Harbor Botanical Garden.

In contrast to the large public gardens, smaller retreats hover high above the bustling city. These are the rooftop gardens, small jewels of flowering plants and their pollinators: hummingbirds, butterflies, moths, and bees. Hot on their heels are the predators that feed on them: dragonflies, damselflies, wasps, and birds.

One such garden was constructed at the *fin de siècle*, the original Waldorf-Astoria roof garden, a gathering place for the hotel's guests on hot summer nights to sit under the stars and imbibe alcoholic and nonalcoholic drinks. In winter, the roof was flooded to make an ice skating rink. It was in use year-round. Rooftop gardens sprouted up all over the city in the days before air-conditioning: on top of entertainment venues such as the Casino Theatre, which opened the first rooftop garden in New York City in 1883, and atop famous apartment buildings, such as the Ansonia Hotel on Broad-

Original Waldorf-Astoria roof garden, 1903. *Waldorf Astoria Archives*

way and West 73rd Street, built in 1904 with a rooftop farm that contained ducks, goats, and 500 chickens. Each morning bellhops delivered fresh eggs to the tenants.

Some years before David Garcelon worked at the Royal York Hotel in Toronto, Maher Hanna was director of engineering there. Hanna was very much into sustainability and the environment. As the chief engineer, he helped install the hotel's original roof garden in the 1990s. His carpenters built the raised beds. David knew of him but had never worked with him directly. Coincidentally, and fortuitously, Hanna came to the Waldorf as director of property operations and was followed some years later by David Garcelon.

When David arrived, he had little interest in developing a kitchen garden or apiary at the Waldorf. "My first week here at the Waldorf I was very happy to be leaving the garden and the bees behind and not think about them or talk about them anymore," he said. "I had done hundreds of garden interviews with the press, and after five years, I could do the interviews in my sleep. It never even came up when I was interviewed here."

The director of property operations in a hotel is important to sustainability in terms of the kind of light bulbs they buy, usage of water and electricity, and other matters. Almost upon arrival Hanna met with the new chef, and they realized their shared passion for pushing the envelope on green hotel management. David recalled:

> Maher had done a number of things here, and he was happy to have an ally. After I was here just a couple of days, he said, "Come with me." We took the elevator to the twentieth floor. We walked around the roof, and I honestly had no interest in putting in a garden. None. It was something I left behind, a closed chapter. I just wanted to be the executive chef at the Waldorf. I had plenty to do. But he walked around with a big smile on his face and said, 'Look at this!' and it was so much better than Toronto. In Toronto, you had to walk up three flights of stairs, and anything you had to take there, in or out, like bags of soil, you had to cart up. On the twentieth floor, here you can walk right out. It's very close to the service elevator. At the time, it was really a mess, but it was a big space, with a beautiful view.

The view, in fact, is spectacular.

Looking south from the garden stands the Chrysler Building. During the day, its silver steel cladding shimmers in the sun, and at night, the lights of its crown sparkle. To the west, you can see the Hudson River, to the east the East River and the tip of the Four Freedoms Park on Roosevelt Island. Up and down Park Avenue stand the towering glass and steel office buildings, with the Seagram Building to the north next to the colorful Byzantine dome of St. Bartholomew's Church.

David Garcelon continued:

> Maher said, "I want to do a garden here. I'll pay for everything out of my budget. We'll build the beds." We talked about it, and I had lunch with my boss, Eric Long. He was the general manager at that time, and he'd been here for over twenty years—a great guy but very intense, and he asked how things were going. I'd been here maybe a month or so. New job, new boss, and we were having this meeting about all there is to do. At the end of the

meeting, he said, "Do you need anything from me?" and I said, "Well, there is one thing. I was talking to Maher, and on the twentieth floor . . ." and before I even finished he said, "You want to put a garden up there." And he's getting up from the table and he says, "No problem. Anything else?" And after my experience in Toronto I kind of swallowed and said, "Also we're thinking about putting some beehives up." He said, "No issues, no issues. Just do it." And he walked away. That was it.

Waldorf-Astoria roof garden ice rink, 1905. *Waldorf Astoria Archives*

Waldorf Astoria Hotel and St. Bartholomew's Church, Park Avenue.
Waldorf Astoria Archives

The Clock, main lobby, Waldorf Astoria.
Waldorf Astoria Archives

Park Avenue Lobby, *Wheel of Life* mosaic.
Waldorf Astoria Archives

Cole Porter piano, Peacock Alley, pianist Emilee Floor.
Leslie Day

Peacock Alley Sunday brunch; left to right: Chefs Charlie
Romano, David Garcelon, Peter Betz, and Christopher Eagle.
Waldorf Astoria Archives

Foraging worker delivering pollen to a pollen cell while returning nectar foragers are waiting to regurgitate nectar to honey processing bees. *Peter Betz*

Honeybees entering and leaving their hive on the Waldorf's roof garden. *Gabe Kirschheimer*

Andrew Coté smoking bees to calm them as Chef David Garcelon looks on.
Gabe Kirschheimer

Honeybee foraging lavender flower, rooftop garden. *Peter Betz*

Chef David Garcelon surrounded by the chefs' garden. *Laura Manske*

Worker bees tending white, C-shaped larvae. *Leslie Day*

Fare Share Friday logo. *Waldorf Astoria Archives*

Waldorf Honey jars. *Beth Bergman*

Fare Share Friday dinner, St. Bartholomew's Church. *Leslie Day*

LIFE OF THE QUEEN

MOTHER OF THEM ALL

> What is true is that a colony's queen lies at the heart of the
> whole operation, for a honeybee colony is an immense family
> consisting of the mother queen and her thousands of progeny.
> — Thomas D. Seeley, *Honeybee Democracy*

She's fairly easy to see because she's the largest bee in the hive. Bearing a long golden or dark brown body, she is surrounded by her daughters who nurture her.

Queens are not born royalty; they are made that way by the workers who feed royal jelly to the chosen few throughout their larval stage. While worker larvae are fed predominantly from their tenders' hypopharyngeal glands, queen larvae are fed a food prepared in the nurse bees' mandibular glands containing much larger quantities of the enzyme biopterin, which contains vitamin B7 and pantothenic acid, also known as vitamin B5. Most important, royal jelly contains the protein royalactin, which causes the queen to grow very large. Royalactin contains hormones essential for the development of ovaries. It also shortens the period of development so that queen larvae go through metamorphosis more quickly than workers and drones. Queen larvae are also fed more sugar in the form of royal jelly. Sugar acts to increase their appetite, ensuring that the queen larvae will eat more. We are similar to bees in that way: the more sugar we eat, the more we want to eat. Sugar is an appetite stimulant.

Being fed with abundant and rich food triggers the larvae's DNA to create an oversized bee with enormous ovaries capable of laying 100,000 eggs or more every year. She keeps the hive filled with her daughters, creates the sons that will seek mates, and even her own heirs, the queens-in-waiting.

As a queen matures, she emits pheromones, powerful chemicals that communicate her needs to her daughters and helps control the social interactions of the hive. From her, the order of the hive emanates. One of her chemicals is known as the queen mandibular pheromone, which attracts large groups of her worker daughters who surround her and attend to her every need. The queen mandibular pheromone, also known as queen substance, is produced in great quantities by a healthy queen. She spreads it around her body when she grooms herself, and it attracts a circle of six to ten daughters who touch her with their antennae, their legs, their mouthparts, feeding her, cleaning her, and removing her feces. After a few minutes of care, these attendants fan out into the hive, spreading her pheromones, which let the tens of thousands of hive mates know that their queen is healthy and strong.

Queens become sexually mature when they are young, typically five or six days after they emerge from their cells. It is at this moment that their retinue of workers start seriously attending them. Life around the queen explodes with activity. The attending worker bees start vibrating against her: pushing her and pulling her. This forced exertion helps prepare her body for mating flights by continuously making her move around the nest. If they are too rough, she emits sounds known as piping, which make the attendants freeze on the spot and stop vibrating. At the same time, the fanning bees get to work at the entrance to the hive and release Nasonov pheromones to help her find her way back to the hive from her mating flights.

Once she is at the entrance to the hive, the young virgin queen may try to flee back inside, but the workers force her back out until she takes flight. She may make a few orientation flights so she knows how to get back to her hive. Eventually, usually in midafternoon, she looks for drones at a congregation site. Over a period of several days to several weeks, depending on weather conditions, the queen makes her mating flights until she obtains the millions of sperm she needs.

The queen stores these sperm inside her spermatheca, an organ connected to her ovaries that holds the sperm in suspended animation for up to several years. With each egg laid, she decides whether to release sperm, thus either fertilizing the egg or not and so producing workers and drones, respectively.

The queen produces fewer pheromones as she ages. This drop signals the workers that it is time to start creating a new queen for the colony. First, they build extra-large and extra-long, peanut-shaped brood cells at the edge of the comb, called queen cups, and start filling them with royal jelly. The queen then lays an egg in each cup. Sometimes a worker bee will move an egg or a small larva into a queen cup, speeding up the process of queen rearing.

On average, it takes fifteen to sixteen days for a virgin queen to go from her egg stage to her adult stage. Once she has mated, she will continuously produce up to 2,000 eggs in a single day.

According to Cornell honeybee scientist Dr. Thomas D. Seeley, right before the old queen's death, when virgin queens start emerging from cells to replace her, the first virgin queen to climb from her cell will move from queen cell to queen cell. Chewing a hole in the side of each cell and inserting her stinger, she stings the queen-in-waiting, killing her. If several virgin queens hatch simultaneously they will fight to the death until the sole survivor reigns as the new queen, the focal point of the hive. Seeley's description in his book *Honeybee Democracy* is both horrific and riveting: "If two or more virgin queens emerge together, they will fight to the death, seizing each other and attempting to sting. The battling queen bees grapple and twist, each one struggling fiercely to plant her venom-laden sting in her sister's abdomen. Ultimately, one queen succeeds and the other, fatally stricken, collapses in paralysis, falls from the comb and soon dies."

BRINGING BEES AND A
GARDEN TO THE WALDORF

Growing mint on rooftops goes way back in New York.
The little urchins of the Five Points used to do it, before the Civil War.
They would grow the mint in rooftop flower boxes, fertilized
by the abundant pig manure to be found on the streets at the time.
They sold it to the best hotels in town, where it was very popular
in refreshing summer drinks. The hoteliers did not,
I'm sure, reveal its origins to the guest.
— Kevin Baker, Author

David Garcelon started the process of creating the honeybee garden on
the roof of the Waldorf by arranging for the bees. He contacted his
beekeeper in Toronto who introduced him to Andrew Coté, the head of
the New York City Beekeepers Association. The chef and the beekeeper met
late in November 2011. They hit it off immediately, and the deal was sealed.
Andrew would help. By April 2012, the Waldorf carpenters and volunteers
had built and installed six beehives. There was a competition among the staff
at the Waldorf to name the hives. The winning names were: Hive of Fame,
the Royal Sweet, the Presidential Sweet, Grand Bee Room, Empire Hive of
Mind, and (my favorite) It's Hive O'Clock Somewhere.

David smiles as he remembers the day that the bees arrived at the
Waldorf.

It was great. We did things a little differently than that time in Toronto. We had a limo pick Andrew up with the bees. We bought six nuc boxes with 5,000 bees and a queen in each box from a breeder to start out. (Nuc is an abbreviation for "nucleus colony." Nuc boxes are used for transporting small colonies with their queens.) We put Andrew and the six hives in a Mercedes limo. Our public relations team was involved, and they let the media know. There were TV crews outside the hotel, some print media, and photographers. We had a dozen chefs and hotel managers out on Park Avenue waiting. All of us: the media, the chefs, the managers, were all waiting on Park Avenue for the arrival of the bees, but the guests of the hotel and people on the sidewalk thought there's some famous guest arriving. So there's this crowd and everyone's pushing. The sidewalk's blocked, people waiting to see who's going to pull up. The limo pulls up and Andrew gets out. He's wearing sunglasses, and he hands the boxes to our chefs Peter and Joshua, and we carry them right through the lobby, right up the stairs, with people taking pictures.

Beehives on the Waldorf Astoria roof. *Leslie Day*

There was no garden yet to feed the bees. So during the warm months of that first year, the bees foraged for nectar and pollen on street tree flowers. Andrew Coté says the bees love the nectar of the silver linden trees, which bloom in late June with pale-yellow and ivory flowers that fill the entire city with a heady perfume. They also prefer the flowers of the Yoshino cherry trees, which have pink blossoms that open in early to midspring. The streets of New York City are lined with a veritable urban forest. Over half a million trees are flowering from late winter through late summer. Some, like the witch hazel tree, flower in both March and December. The red maple tree flowers in late winter to early spring. The honeybees are not active in winter, but once the temperatures start to rise above 50 degrees Fahrenheit, they begin foraging for nectar and pollen. The bees will even travel a mile away to forage on flowers in Central Park.

Spring and early summer blooms give New York honey a light and minty flavor. In autumn, the honey is darker, as the bees forage on late-blooming flowers like aster, Japanese knotweed, and goldenrod.

> I dragged my feet on putting in plant beds, and Maher kept pressuring me. I dragged my feet because I wanted a partner. I wanted someone like we had in Toronto to help with the garden. I knew enough to know what I didn't know, and I'm not the person to teach someone how to garden and do it well. It's like cooking: you need to do something for years to do it well and if we're going to do something here at the Waldorf we need to do it really well. I reached out through contacts and I got in touch with someone at the New York Botanical Garden and they sent somebody. He came right away, and I told him what I wanted to do and I never heard from him again. Quite a bit of time went by and our carpenters built the beds identical to the ones we had in Toronto. Eventually, I got in touch with the Horticultural Society of New York, and George Pisegna came over and was enthusiastic. He said they had all the resources to put towards it. "You guys pay for the materials and we'll do the work and give you our expertise." That got the garden moving again.

When the raised garden beds were first installed, David had a lot of volunteers: kitchen staff, salespeople, and members of the Waldorf's sustain-

Building new garden beds.
Leslie Day

ability committee, which included both David and Maher Hanna. David explained the committee's functions:

> We do a ton of projects from collecting soap that guests don't use, to reducing water consumption, to recycling, to community service. We asked for volunteers and on two or three Saturdays people came in and painted the raised wooden beds that the carpenters had built. Together we all put the soil in. George told us what we needed for the soil mixture and we got the beds ready. George came with some people being trained by the Horticultural Society. They have a program with the New York Department of Corrections where they have low risk offenders that they teach to be gardeners and they brought them to care for the plants.

The Horticultural Society of New York does groundbreaking work, literally, with former detainees and inmates from Rikers Island Correctional

Facility and other New York prisons. Once the men and women leave prison, they are offered opportunities for job training and support in several fields, one of which is horticulture. The Horticultural Society of New York, informally known as The Hort, has an internship program that helps former detainees and inmates who participated in their GreenHouse program in horticulture on Rikers Island, run by Hilda Krus. Hilda is a trained horticultural therapist who is known as Saint Hilda by her colleagues at The Hort. She travels from northern Manhattan, where she lives, to Rikers Island each weekday to work with detainees and inmates in the prison island's gardens and greenhouses. As part of the transition back into their communities, the released men and women continue their horticultural work. The Hort's website states that the "GreenTeam provides vocational training in horticulture, transitional work, job search skills, job placement, and aftercare services."

David reflected on the garden as we sat and talked that winter day and started thinking about spring planting. "One of the important roles that the garden is playing is that in winter we ask ourselves, what are we going to grow this year? What did we grow last year that was useful?" When I asked him who "we" was, he said the chefs.

There are 140 chefs at the Waldorf, but there are only a handful involved in choosing what's grown in the garden, including Peter Betz, and Charlie Romano, the pastry chef. So I asked them what they liked having in the garden; what they used, what they didn't use, if they need anything new. To me it's all about the story—the story is important. Bees are a wonderful story. It's honey from New York City, that's cool; from this hotel, that's even cooler. It's from the roof and our chefs harvest it. That's a great story. With the garden, if we grow parsley on the roof and chop it up and put it in salad, no one cares very much; but if we're growing black tomatoes, or some cool variety of apples or some flower that no one's ever heard of—people notice and they care. I'm trying to get George Pisegna to get me a paw paw tree. Paw paw is a really cool fruit that no one knows about. It's native to this part of North America. Paw paws almost taste like a tropical fruit but you can't buy them, and very few people know about them. I want to grow paw paws here so we can use them, and so we can tell our customers, "Hey, we grow these paw paws on the roof!"

David appreciates the intricate details of where food comes from, how it is used, and how it makes the person tasting it feel.

> People always ask me, why? Why do you have a garden? I think there are several reasons. First, it's for my chefs. Although I grew up never knowing much about gardening, I now think it's important for us to know where our food comes from. We are becoming disconnected from the seasons and our local food, and I've seen it getting worse. There is an expectation now with our customers that we'll have strawberries and blueberries in January. People go to a farmer's market in New York in May and see bell peppers and think they're local—they're not—no way. As chefs we need to be experts, not experts like farmers are, but we need to know more than the average person about where our food comes from and what grows when and where and get excited about it.

The once reluctant David is now thrilled when he and the chefs go up to the garden during mid-June and collect zucchini blossoms for a pasta dish.

David's face lights up when he talks about his favorite herb growing on the roof.

> Whenever I take chefs or people that are into food or cooking up to the garden, I always take them over to the tarragon. I walk over to the bed, peel off a leaf and say, "Taste this," and they are amazed. It's better than anything that we can buy because it is fresh. For some things, like thyme or rosemary, maybe it doesn't matter that much, but with tarragon it matters a great deal, and when you're a chef and you taste this, you say, "Wow!" and you start to think of tarragon differently; it's not just another herb you get from the fridge. You think that this is the best tarragon you've ever tasted—and you say to yourself, "What can I do with that? Let's make a great *Béarnaise* sauce for the chicken."

Waldorf Astoria chefs on the rooftop garden, 2014. *Jonathan Stas*

I asked David whether this is the real reason to grow your own: if it's local, it's fresher; if it's fresher, it tastes better and is probably healthier. David's answer was more than I expected:

> The thing about food is that it's complicated. Everyone looks for the headline—don't eat this, eat that, it's healthier; only eat organic; organic is better. But it's not that clear-cut. It's gray. Some things are better organic, some things are not; some things are worth getting organic, some things are not. Some things are better for the environment if they're organic, some things are not. Should you buy organic apples? Maybe. I don't know about New York State, but in Ontario, it's almost impossible to grow apples organically because you need pesticides to keep the apples free of the many things that love to eat them. Is that a terrible thing? Not if it's done responsibly. Should we not grow apples in

North America? They are not native to North America in the first place—so should we not grow them at all? I don't think that's the answer. Should we be flying in organic apples from New Zealand or Chile? Because a lot of time when you buy organic apples, that's where they are coming from. Is that the best thing for the environment: take a case of apples that weighs fifty pounds and ship it halfway around the world? I'm not saying organic is bad, I'm saying, let's ask questions about it. Where are these apples coming from? How are they growing them? So the garden is educational for my chefs, first and foremost.

The garden is also a statement from the hotel that it is trying to do something good for the environment. David thinks that every chef and every restaurant should do something.

I don't care if it's an old oil bucket by your back door with some mint growing in it. Everyone can do something. We are lucky with our resources: we have this great space and carpenters and people in the hotel who volunteer, and this relationship with the beekeeper and the Horticultural Society, so we can do more than most, and we should do that; that's part of being a good citizen, a good member of the community. One of the things that drives our sustainability committee is that we're an eighty-year-old hotel with all kinds of challenges and infrastructure issues, and if we can do things to make the hotel sustainable then maybe anybody can. We want to set a good example.

BEEKEEPING THROUGH THE AGES

> One can no more approach people without love than one
> can approach bees without care. Such is the quality of bees.
> — Leo Tolstoy

Winter can come suddenly and with a vengeance in New York City. When snow covers the honeybee hives on the Waldorf Astoria's roof, Andrew Coté, New York City's humble chief beekeeper, checks to make sure all is well. With soft mounds of white over the gardens and hives, he feels the chill of snow as the wind occasionally gusts twenty stories above Park Avenue. Most of us would assume the bees are dormant, but Andrew knows differently. They are awake and busy.

He's come in the aftermath of a storm to check on them. His first task is to make sure no lids have blown off the hives. Then he checks the entrance to each hive, pushing snow away so that the airflow is not blocked. Trudging from hive to hive, he is eventually satisfied that all is well.

New Yorkers are as fascinated by Andrew as he is by bees. The media often run stories about him, and describe how four generations of his family's passion has become his life's work. Growing up in Norwalk, Connecticut, Andrew's father was a firefighter but continued the family tradition of managing up to sixty apiaries in the region surrounding their town: on working farms, horse farms, and fields. His father's family is French Canadian and has been raising bees since the 1880s.

Andrew remembers traveling with his father to check on the hives as a small child. He helped by carrying the smoker, the bee suit, long gloves, and the hive tool, which is used to chisel open the lid and frames. I asked him

Honeybee garden in snow, beehives wrapped for winter. *David Garcelon*

whether he was ever afraid of being stung and he quoted his father, who often said, Andrew "lacked the common sense to ever be afraid of anything." He remembers reaching his young hand into the hives to scrape out the soft, warm honey.

Throughout the honey season, he and his father sell their honey at the greenmarkets throughout New York City, including the Union Square Greenmarket on Wednesdays and Saturdays, from a stall surrounded by a large, white canopy bearing a sign that reads "Andrew's Honey, Taste Bud-Bursting Local Wildflower Honey." On the days when he is not selling his honey, he is checking his seventy-five hives scattered throughout the city, including at the High Line and the Waldorf Astoria, as well as rooftops on the Lower East Side, the Upper West Side, Williamsburg and Greenpoint, Brooklyn, and Forest Hills, Queens. When Andrew and his father sell their honey at the local greenmarkets, the jars are labeled with the locations. On Andrew's honey jars, you will see "Bushwick," "Coney Island," and "Tribeca" among the labels of the dozens of beehives that he manages in the city.

Andrew is president and founder of the New York City Beekeepers Association. In 1999, beekeeping was listed as an illegal activity in New York City. Honeybees were added to a list of dangerous animals that included lions, tigers, bears, hyenas, tarantulas, cobras, and dingoes, among others. Then, in 2010, under Mayor Michael R. Bloomberg, the city's Department of Health, in keeping a ban on most venomous insects, lifted it on honeybees:

> Persons keeping honey bees shall file a notice with the Department, on a form provided or approved by the Department, containing the beekeeper's name, address, telephone, e-mail and fax numbers, emergency contact information, and location of the hive, and they shall notify the Department within ten business days of any changes to such information. Beekeepers shall adhere to appropriate beekeeping practices including maintaining bee colonies in moveable-frame hives that are kept in sound and usable condition; providing a constant and adequate water source; locating hives so that the movement of bees does not become an animal nuisance, as defined in §161.02 of this Article; and shall be able to respond immediately to control bee swarms and to remediate nuisance conditions.

Beekeeping was once again legal, and since then hives have popped up in all five boroughs: in backyards; atop restaurants, hotels, and office buildings; in city parks; on fire escapes; and on apartment rooftops and terraces. In New York City, there are now nearly 300 registered beekeepers. The most elevated beehives, a collection maintained by Andrew Coté, are atop the Marriott Residence Inn near Central Park on Broadway. The beehives are kept on the roof of the seventy-sixth floor, making it the highest apiary in the world.

For centuries, beekeepers have tinkered with beehive designs, always looking for ways to extract the sweet, golden honey without killing the bees and destroying their colony. During the thousands of years of early beekeeping, hives were often simple clay cylinders in which bees built their combs along the hive walls. The bees were driven off the honeycombs with smoke, and then the combs were cut out of the hives and pressed to squeeze out the honey.

The long, cylinder-shaped hives made with mud, straw, and dung are still in use in some places in Egypt and elsewhere. Straw hives, known as skeps, represent a different design and have been in use for more than 1,600 years in Europe. When the Saxons invaded Britain around AD 400, they brought skeps with them. The word *skep* is derived from the Norse word *skeppa*—a straw basket to measure grain. Skeps, made from long lengths of dried stalks of grain plants (oats, rye, wheat, barley, and rice), were made by skeppers who used a leather ring or the horn of a cow (called a *girth*) to control the thickness of the straw coils as they pulled them through. The coils were sewn together with flexible briar stems sliced into thin strips after the thorns and prickles were removed.

From the British Isles, early colonists migrated to the Americas and brought along their straw skeps. With few experienced skeppers living in America, but with an abundance of trees and lumber, there was a shift from straw skeps to hives made of wood. Early settlers also used sections of bee trees. They cut out the sections of trees that contained colonies of bees, often from black tupelo trees (*Nyssa sylvatica*, also known as gum trees), whose flowers produce a rich source of nectar for bees.

Desiring a source of honey and beeswax, the American colonists had honeybees shipped from England—first to the Colony of Virginia in 1622 and then to Massachusetts in the 1630s. Honeybees were in Connecticut by 1644, in New York by 1670, in Pennsylvania by 1698, and then spread to southern states and the Midwest in the eighteenth century. In the 1850s, honeybees were shipped from the East Coast to California and eventually made their way to Oregon, Washington, and the mountain states, either as wild swarms or because pioneers transported them. Along with native wild bees, pioneers now had honeybees to pollinate their thriving crops and orchards. Throughout the early history of beekeeping in America, obtaining the honey often meant destroying the comb and killing bees.

As a child in the 1820s, Lorenzo Lorraine Langstroth was a passionate student of social insects. He wore out the knees of his pants observing the movement of ants as they traveled along the ground in front of his childhood home in Philadelphia. After graduating with a degree in theology from Yale in 1831, he visited a friend who kept bees and had a large glass bowl in his parlor filled with honeycomb. The friend took him up to his attic where he kept his beehives. Langstroth was enamored by the bees and bought two hives on his way home, eventually moving his new passion to Philadelphia when he returned to his boyhood home. Langstroth suffered from serious bouts of depression and chronic headaches, which affected his work as a minister and teacher. Studying bees helped lift him out of his pain and his mood swings. He went from owning two log hives to maintaining hundreds and found himself spending every spare moment learning how to care for and observe his bee colonies. He established a large apiary and set out to create a hive where honey could easily be removed without harming the bees.

The middle of the nineteenth century was a difficult time for beekeepers because of the larvae of the wax moth, a pest that fed on the beeswax combs and covered hives with their sticky webs and cocoon shells, which rotted. At the same time, a deadly and contagious bacterial infection, American foulbrood, was killing bee larvae. There was no way for beekeepers to remove infected bees or to pick out the wax moths without cutting the comb from the hive, killing some bees.

Langstroth was determined to design a beehive that could be inspected without doing damage to the honeybee colony's nest. He read books about beekeeping from the ancients Aristotle and Virgil, who were avid observers of honeybees, to more contemporary European beekeepers. He read *The Feminine Monarchie* by England's Father of Beekeeping, Charles Butler, who was among the first to state that the large bee controlling the hive was a female—the queen. He read works by the Dutch biologist and scientific illustrator Jan Swammerdam, who, in the seventeenth century, spent every minute of daylight examining honeybees. His sweeping compendium *Historia Insectorum Generalis* includes clear and beautiful drawings of bee anatomy. Langstroth wrote, "the interior of a hive was to common observers a profound mystery," and this ignorance, he believed, compounded beekeepers' problems.

Langstroth continued to read accounts of all of the honeybee research and experiments that had gone on before him. He studied the work of the eighteenth-century blind French scientist François Huber, who, aided by his wife and his devoted servant, François Burnens, experimented on honeybees and, as a result, designed a new type of hive, which he called a *leaf hive* that opened like the pages (or leaves) of a book. This allowed observation of the interior of the hive without destroying it. He read the writings of Edward Bevan, the British physician and beekeeper, who in the 1820s designed a hive with stacked boxes. And he investigated the work of Johann Dzierzon, a Silesian pastor, who built a hive that had removable honeycombs supported by grooves in the hives' walls. Despite theft, fire, flood, and foulbrood, Dzierzon had almost 400 hives that produced 6,000 pounds of honey annually.

Even with all of these innovations, it was still not possible to properly care for the bees and remove the honey without breaking the combs and thus damaging the colony's nest. Langstroth wanted to create a better hive, without which beekeepers would be "unable to remedy many of the perplexing casualties to which bee-keeping is liable." To do this, he knew that he needed to design a hive that would not allow bees to attach their comb to the top, bottom, and sides, which was what made it impossible to remove a comb without destroying it. After years of observation, he had an aha moment: bees needed a space, slightly less than three-eighths of an inch to be exact, for a single bee to move between combs. If the space was narrower, the bees would fill it up with propolis: the resinous material collected from tree buds that they used to caulk small spaces. If the space was wider than three-eighths of an inch, they would fill it with honeycomb. This came to be known as *bee space*. Langstroth set out to create a hive that would incorporate the bee space throughout, and, in 1851, he had an epiphany:

> Returning late in the afternoon from the apiary which I had established some two miles from my city home, and pondering, as I have so often done before, how I could get rid of the disagreeable necessity of cutting the attachments of the combs from the walls of the hives . . . the almost self-evident idea of using the same bee-space . . . came into my mind and in a moment the suspended movable frames, kept at a suitable distance from each other and the case containing them, came into being. Seeing by intuition, as it were, the end from the beginning, I could scarcely refrain from shouting out my "Eureka" in the open streets.

Langstroth built his movable-frame hive in 1852. His design is still used by beekeepers today, 166 years later. For the first time, beekeepers could remove and inspect the combs. They could remove and destroy moths and their larvae and other invaders. They could harvest the surplus honey without hurting the comb, the adult bees, or their brood. "Many persons have been unable to suppress their astonishment," he boasted, "as they have seen me opening hive after hive, removing the combs covered with bees, and shaking them off in front of the hives; forming new swarms, exhibiting the queen, transferring the bees with all their stores to another hive; and in short, dealing with them as if they were as harmless as flies." He wrote and published *The Hive and the Honey-Bee* in 1853, describing the benefits of his hive design and dispensing practical advice on beekeeping. The book is still in print.

After thousands of years of beekeeping, Langstroth's invention transformed beekeeping and allowed beekeepers to enter into modern agriculture. His breakthrough inspired other inventors, who helped beekeeping become more commercially successful. Within twenty years of Langstroth's breakthrough, there were three more major inventions.

In 1857, Johannes Mehring, a German master carpenter and beekeeper, invented sheets of beeswax imprinted with the honeycomb pattern inserted in the movable frames used in the Langstroth hive. These premade midribs of honeycombs gave bees a head start, allowing them to focus more of their energy on making honey rather than beeswax. The same forms are still in use today.

In 1865, an Austrian beekeeper named Francesco de Hruschka invented the honey extractor, which used the centrifugal force of a drum spinning inside a barrel to pull honey from cells without destroying the comb. The honey extractor is still in use today.

In 1873, the last of the three major beekeeping inventions was made by a beekeeper living in Upstate New York, Moses Quinby, who devised a bee smoker that added a bellows to the traditional firebox used to calm bees when working in hives.

To let the thousands of American beekeepers know about these new inventions and methods, two national publications were launched: the *American Bee Journal*, which began publishing in 1861, and A. I. Root's *Gleanings in Bee Culture: Or How to Realize the Most Money with the Smallest Expenditure of Capital and Labor in the Care of Bees, Rationally Considered*, which was launched in 1873 and is still the nation's predominant beekeeping journal (though now succinctly called *Bee Culture*).

By the end of the 1800s, it was possible for beekeepers to quit their day jobs and go into beekeeping full time. Quinby developed not only the bellows smoker but also his own variation of Langstroth's hive and was probably the first person to make full-time beekeeping a monetary success, thus earning him the title of "America's Father of Practical Beekeeping." He kept 1,200 colonies in Upstate New York and developed methods for treating foulbrood disease. He was one of the first beekeepers to put supers—extra boxes—on top of his hives so the bees could store large amounts of honey.

More and more apiarists joined Quinby, and beekeeping became a huge commercial success. Like the Egyptians who moved their hives up and down the Nile, pollinating flowers and crops along the shore, American beekeepers move hives from one part of the country to another to pollinate farmers' fields and orchards. Innovations in beekeeping have not ended; for example, the top-bar beehive is gaining in popularity. But the major early advances transformed beekeeping into the system we see today.

During late autumn and throughout winter, Andrew watches the bees at home in his specially made observation hive. Through a lifetime of studying bees, he has a pretty good idea of what the bees are doing when the weather turns cold on the Waldorf Astoria Hotel's roof. When I asked him how the bees survive the cold, Andrew was quick with an analogy. He explained that the bees huddle around their queen in the hive the way penguins huddle to survive in the Antarctic. Penguins on the outside of the circle block the wind and keep the penguins on the inside of the huddle warm. The wind blockers push into the center when they need to warm up, and the warmed penguins in the center shuffle to the outside. In the hive, he explains, the cold bees on the outside will move toward the center, and the warm bees, like the penguins, shift to the edge. But this is where the penguin analogy ends. It's more complicated, and a bit more brutal, inside the hive.

As October wanes, the hive prepares in earnest for winter. With the late September–early October renaissance of fall-blooming flowers behind them, honey production is finished. If the hive is to last until spring, the worker bees will have stored about one hundred pounds of honey. The first step in the preparation is to rid the hive of bees whose functions are no longer needed. Well before the first snow, as autumn's temperatures drop and daytime highs reach only into the fifties, the drones are corralled to the bottom of the hive and cast out to die. With fewer mouths to feed, the hive has a better chance to make it until spring.

The worker bees have another important task to perform to protect their hives for winter. From trees that line the city's streets, they gather resin to make propolis, a bee specialty used to caulk any cracks or holes in the wooden hives. Resin is a thick, sticky substance produced by special plant cells. When a plant is injured, resin is secreted from the wound, sealing it off from the air and protecting it from invading bacteria, viruses, fungi, and insects.

Amber is fossilized resin. Tree bark, flower buds, and leaf buds can secrete resin when extra protection is needed. The bees have adapted its use to protect their homes. Worker bees prepare for winter by using their forelegs to gather resin from the trees and shrubs that line the city's streets, using their mandibles to chew bits of resin from the plant. They then pack the resin into their corbiculae, the baskets on their hind legs. Each corbicula can hold about 10 milligrams of resin, which they carry back to the hive. There they are met by other worker bees that scrape the sticky resin from the foragers' corbiculae, chew it, and mix it with saliva. The end product is propolis, a word allegedly coined by Aristotle that literally means "before the city." Another term for it is bee glue. Propolis is their caulking, used to seal cracks or holes to keep winter's wind and trespassers out of their nests, whether in tree cavities or in human-made wooden hives.

Throughout time, humans have taken propolis from hives. The Romans used propolis to treat wounds and infections. In the seventeenth century, Italian violin maker Antonio Stradivari added propolis to the varnish he used on his instruments. The Greeks, Egyptians, and Assyrians harvested propolis from their beehives and used it medicinally to treat wounds, fight infections, and embalm their dead. Bees use propolis to embalm intruders, such as mice and large wasps, that are too heavy to drag out of the hive. Propolis prevents decay and putrefaction. Keeping bacteria and animal intruders out of the hive is essential, which is why honeybees work so hard to make sure they have a good supply of propolis that will last throughout the long winter. Many little creatures seek out honeybee hives in winter where they will be safe, warm, and well fed by honey.

Andrew helps the bees protect their hives further by placing a small metal guard in front, which reduces the draft and stops honey marauders and other intruders seeking shelter from the plummeting temperatures. As winter sets in and the temperature drops, the next step the bees take to survive is extraordinary. The bees turn into miniature heaters, quickly contracting and expanding their powerful thoracic flight muscles, which are capable of 230 wingbeats per second. The bees can generate so much heat through this feat that their internal temperature can rise to more than 110 degrees Fahrenheit.

A mass of thousands of shivering bees surrounds the hive's queen, who bathes in air that is kept at a constant 92 degrees Fahrenheit. When the bees on the outside of the huddle start to become cold, they take some honey for themselves, their queen, and their sister bees into the center of the huddle. Thus, the constant flow of bees from outside to inside and vice versa allows for the flow of food as well as the warming and cooling of the workers.

As the stored honey in one spot is used up, the bee huddle moves up the frame so that they are always near their food source. When the temperature outside drops, the huddle grows tighter and tighter, reducing the surface area and maintaining the queen at the perfect temperature. If the temperature begins to warm up too much, the bees in the huddle move apart, allowing air to move in and moderate the temperature.

Satisfied that the hives are intact and the queen is warm, Andrew heads back inside the hotel after taking one last look at the hives and the holiday lights below, sparkling on every tree up and down Park Avenue. Looking south, he sees the arches of the jewel-like crown of the Chrysler Building glittering in the cold winter air. As Andrew opens the door and steps inside, the hotel's warm air encircles him, and for a moment he's like the warm bees in the hives. With the next New York snowfall, he'll be back atop the Waldorf Astoria, but for now, he has many other places to go, many more hives to examine.

A MATCH MADE IN HEAVEN

ST. BARTHOLOMEW'S EPISCOPAL CHURCH AND THE WALDORF ASTORIA HOTEL

> I am indeed very much pleased with the doors, as well as with the
> tympanums and the work on the portico at St. Bartholomew's,
> and I am sure it must be the opinion of everyone who sees the
> completed work that it is very beautiful and appropriate
> and that it will mark an era in American Art.
> —Alice Vanderbilt to architect Stanford White, October, 1903

Sitting across the street from the luxury of the Waldorf Astoria is Saint Bartholomew's Episcopal Church. Founded in 1835, it was originally on Lafayette Street, named by John Jacob Astor for the Revolutionary War General Marquis de Lafayette. The church moved to Madison Avenue and 44th Street in 1872 and, in 1914, moved to its current location on East 50th Street between Park and Lexington Avenues. It sits on the site of the old Schaefer Brewing Company, from which the property was purchased. Although the first services were held in the new church in 1918, construction wasn't completed until 1930, a year before its neighbor, the new Waldorf-Astoria Hotel, welcomed its guests.

You can't help but feel a sense of elation when you walk through the glorious doors. The three doors are part of a French Romanesque Revival portal that was designed by Stanford White in 1902 for the prior Madison Avenue church and commissioned by the family of Cornelius Vanderbilt II as his memorial. In 1903, David Greer left St. Barts to become bishop of the

Episcopalian Diocese of New York City, and Leighton Parks became the new rector. It was Parks who determined that the church had to move in 1914 because of the sinking of its foundation on Madison Avenue and the resulting unsound structural conditions. It seems that the land beneath the church was made of soft clay, not the solid bedrock of Manhattan schist that much of New York City buildings are anchored into. The doors were saved, and the entire portal was transferred when the church moved. The bronze doors are richly sculpted with panels of bas-relief portraying scenes from the Old and New Testaments. They are bordered with foliage carvings of acanthus, oak, ivy, thistle, and laurel leaves. Images of a dragon, fowl, pelican, peacock, and eagle add to the scene.

Cornelius Vanderbilt II was actively involved in St. Bartholomew's Church. As a young man, he taught Sunday school at the church, which is where he met his future wife, Alice Gynne, who also taught Sunday school at St. Bart's. Throughout his life, Vanderbilt was a generous philanthropist, and St. Bart's was often the beneficiary. In 1891, he and his mother, Maria Louisa Vanderbilt, built a parish house for St. Bart's on East 42nd Street,

St. Bartholomew's Church, Central Portal. *Library of Congress Online Catalogue*

which served the poor and immigrants who were coming to the city in a tidal wave. Throughout the late nineteenth century, the Vanderbilts gave financial support to the church, helping Rector David H. Greer, who led the church in its fight against elitism and toward the goal of serving the poor and lower middle class through social services. With financial support from Vanderbilt, Greer opened six Sunday schools in five languages, a rescue mission, and the Parish House and Medical Clinic on 42nd Street.

With the death of Vanderbilt at the age of fifty-five, his wife, Alice, a devotee of the church, wrote to Greer that she and her daughter, Gertrude Whitney, were "desirous of putting in St. Bartholomew's Church something as a memorial of my husband. We have thought of bronze doors at the church entrance." Greer reached out to the famous architect Stanford White, who had designed work for other members of the Vanderbilt family, though not their 137-room mansion on 5th Avenue between 57th and 58th Streets or their still existing 70-room Newport, Rhode Island, summer "cottage," called The Breakers.

Years before, as a young man, Stanford White had seen doors similar to what Mrs. Vanderbilt wanted, in a small church in St. Gilles, France, while on a summer trip with his friends, the architect Charles McKim and the sculptor Augustus Saint-Gaudens. When he was hired to design the portal, White modeled it after the portal to the Provençal Romanesque church, Saint-Gilles-du-Gard near Arles, France.

St. Bart's is a jeweled building created by architect Bertram Grosvenor Goodhue in the Byzantine Revival style. It is topped with an eight-sided dome covered in colorful terra cotta tiles, marble, and granite, and capped with a gilded cross. The dome was added to the church after the death of Goodhue. As you step inside the narthex and look up, you find the art deco work of artist Hildreth Meière. Her golden dome mosaics depict the six days of creation, made stunning by her use of glass and 24-karat gold leaf that employed hundreds of shades of gold. The church's stained glass windows surrounding the nave are also by Meière as is the brilliantly colored Hispano-Moresque Transfiguration mosaic of glass and gold leaf above the apse.

Everywhere you look in the vast space within the nave you see surfaces covered by gorgeous colored glass, mosaic tesserae, marble, wooden paneling, and carved stone bas-relief. Adding to the awe is the Aeolian-Skinner symphonic organ, the largest church organ in New York City. More than

Art deco artist Hildreth Meière's 24-karat gold leaf and glass mosaic domes, narthex, St. Bartholomew's Church. *Leslie Day*

12,000 pipes are located in three main parts of the church, including the celestial organ, whose pipes are behind the dome's decorative woodwork directly over the nave. Looking into the massive barrel-vaulted nave you see, adjacent to the dome, intricate and interlaced wooden squinches that cover the highest point of the organ's pipes. St. Bart's has been a leader in the culture of religious music for more than a century. By 1900, it had a full choir of men and women, and, in 1905, Leighton Parks and the church hired conductor Leopold Stokowski to lead the choir.

There is something about St. Bart's, something deeper than any meaning you might find in the glorious features of the building: it is the charity of the church's clergy, leadership, and members. For 130 years, the church has tried to alleviate the suffering of the disenfranchised. Today, it takes the organized form of a group called Crossroad Community Services.

Perched directly across the street from the opulence of the Waldorf Astoria, St. Bart's could cater exclusively to Manhattan's rich and famous. Instead, it continues to open its arms wide to all, the rich and the poor and everyone whose fortunes fall between those economic extremes. St. Bart's, there in the middle of Manhattan, is one of New York's leading charities. They welcome in those they believe God has called them to serve. One of the people they also would welcome was Chef David Garcelon and his colleagues at the Waldorf.

14

IN THE DARK OF THE HIVE

THE SENSORY WORLD OF THE HONEYBEE

> For the bees, the flower is the fountain of life;
> For flowers, the bee is the messenger of love.
> And to both, bee and flower, the giving and the
> receiving of pleasure is a need and an ecstasy.
> —Kahlil Gibran

The bees' pressing need for food is fulfilled by the flowers that await them. And whether they visit for pollen or nectar, they do the job that flowers want them to do: carry their pollen, rich with sperm cells, from one flower and deposit it on another of the same species, so the flower can create seeds and reproduce.

The life of the colony depends on workers finding flowers and bringing back the nutrient-rich provisions, nectar and pollen, to nourish their community. Thousands of bees leave the hive each day to harvest these riches. The entire operation requires a variety of senses, some of which are used to communicate the location of floral resources.

When the first generation of the Waldorf's honeybees was brought to the roof, the flower garden was still in its planning stages. The majority of first-foraging flights almost certainly went from their home at 49th Street and Lexington Avenue to Central Park, about a mile away. There they harvested from abundant flowers in the famous urban oasis. Some of the bees would have stopped short of Central Park to reap the offerings from flowers on street trees and trees on the planted islands of Park Avenue. Others likely found flowers on terraces in Midtown Manhattan.

In so many ways, the sensory world of honeybees is different from ours. They use the combination of scent, landmarks, direction of the sun, electrical fields, wind speed, ultraviolet light, and the communication dance of returning foragers to help locate flowers.

Odors play a primary role. Foragers return to the hive carrying the perfumed scents of flowers that cling to the hair that covers their bodies. A bee in the hive exposed to the scent found on a successful forager is able to set out and recognize the same species of flower in a garden.

In 1901, the Belgian essayist, writer, and Nobel Prize winner Maurice Maeterlinck published *The Life of the Bee.* He had been studying honeybees for years and started experimenting to further understand how bees know where to find nectar. He placed a dish with sugary syrup outside and followed a marked bee that had visited it back to the hive. He let the bee enter the hive but caught her as she exited to return to the dish. However, there soon appeared a large number of other bees at the syrup, without being led by the original, now captured, bee. Maeterlinck surmised that somehow the returning bee had communicated the whereabouts of the sugar source to her sisters in the hive.

Karl von Frisch, an Austrian animal behaviorist and fellow Nobel laureate, would reveal the mechanism of communication. Employing careful experiments, he determined that the dance of the honeybees was a form of language. The waggle dance of returning bees, he found, indicates the flower's direction relative to the sun and their distance from the hive. Within the darkness of the hive, the bees press close to the dancer and, using their marvelous senses, learn the direction and distance indicated by her dance. They are also in tune to the duration of the dance: the better the floral source, the longer the dance advertising it. They use their antennae to "hear" the vibrations of the dancing bee's wings, and their legs to feel the shaking of her body through the vibrations in the comb.

When they find these flowers and start foraging, honeybees use their senses to get information directly from the flowers. Remarkably they can sense and change the electrical charge of flowers. Honeybees have a positive charge. Unvisited flowers are negatively charged, but when a bee collects pollen and nectar from a flower, the electrical charge of the flower changes from negative to positive. When other bees approach the same flower shortly afterward, they detect the positive charge of the flower and recognize that its treasures have been taken. Instead of wasting time, they move along to find a negatively charged flower.

Some flowers let a bee know when it has no more pollen by changing its colors. *Aster divaricatus*, commonly known as white wood aster, a widespread flower of late summer and autumn, has white ray flowers surrounding a central disk of tiny yellow florets, which contain both male and female parts, offering both nectar and pollen. These yellow florets turn red when they are low on pollen supplies, signaling to honeybees and other pollinators to move on to flowers with yellow florets. This saves energy on the part of the pollinator and is a method the flower uses to control the behavior of the pollinating insect, which it needs to move its pollen.

Pollen (from the Greek *palynos* for dust) grains are produced by the flower's anther, part of the stamen, which is the male reproductive organ of the flower. The outer cover of the pollen is the exine, which is highly sculptured with striking colors and beautiful patterns characteristic of each plant's species. Made up of spines, ridges, depressions, and pores, the pollen surface is extremely interesting to look at under a microscope. The pollen grains of bee-pollinated flower species are not only colorful, spiny, and highly sculpted but also sticky. All of these features help them cling to the hairs on the mouthparts, legs, eyes, and bodies of the honeybees that carry thousands of pollen grains back to their hive.

Inside the pollen grains are several cells. One of these cells, the germ cell, divides into two sperm cells, which carry the male DNA of the plant. Another cell develops into the pollen tube. Pollen grains have pores in their walls, which, once they have landed on the stigma (the tip of the flower's female organ, the pistil), allow for the germination and growth of the pollen tube out of the grain into the stigma, down through the style, into the ovary, and finally into the ovules (eggs) that contain the female DNA of the plant. The sperm cells then move down the pollen tube into the ovules, fertilizing them, causing them to go from egg cells to seeds that hold the embryo, the plant's next generation. Unless they are wind pollinated, flowering plants completely depend upon animal pollinators, those "messengers of love," to help them complete their sex act.

Not easily destroyed and with characteristic surface ornamentation for each plant species, the science of pollen analysis has been used to date pollen found in archaeological digs, to cite the floral provenance of honey, and in police forensic investigations, where the clothing, skin, and hair of victims are searched for pollen grains, to see what plants were nearby.

Pollen has evolved as a necessity for bees and has come to be more attractive, that is, more edible and digestible for them. It is their major source of protein. Pollen grains can be microscopic — so tiny that one flower or inflorescence (one flower on a stalk of flowers) can produce tens of millions of pollen grains. Pollen also contains other nutrients: minerals, vitamins, sugar, and starches that are essential nutrients for bees. Flowering plants and bees evolved together over millions of years and have a symbiotic relationship. Flowering plants need bees in order to reproduce, and bees need the nectar and pollen that flowers produce in order to live.

There is a village in China that made its living from an apple orchard, but the overuse of pesticides and removal of wildflowers that their bees foraged from wiped out the local bees. To pollinate the apple flowers, every man, woman and child had to pollinate the flowers by hand. That spring when the apple trees bloomed, children stopped going to school and their parents could not work at other jobs. Can you imagine what our world would be like if we had to hand-pollinate all of our crops?

Forager bees carry pollen back to the hive where they kick their pollen loads off into empty cells, and then the pantry bees working inside the hive process it for storage. These workers add enzymes and honey to the pollen when it is packed into storage cells to prevent fermentation. After this processing, the pollen is called bee bread and is now ready to be digested by larvae and adults, including nurse bees, who need good nutrition to care for the thousands of larvae in the hive.

The most important parts of the bee's body for collecting pollen are its body hairs and its six legs. To understand how the honeybee carries pollen and propolis back to the hive, you have to look closely at the structure of their legs. A honeybee, like all insects, has six jointed legs attached to its middle body section, the thorax. These legs, used for walking, are also essential tools for grooming the hair covering their bodies, for collecting pollen and propolis, and for packing these resources into pollen baskets to carry them back to the hive.

Their front legs are also used to clean the bees' antennae. Built into these first pair of legs are antennae cleaners in the form of a curved notch with stiff bristle hairs and a nearby spur through which they pull their antennae, wiping them clean. The bee does this by raising her foreleg over her antenna and bending the last segment of her leg, called the tarsus, which closes the notch, forming a ring around the antenna. She then pulls each antenna through the bristles to brush off the debris. Observers have seen bees do this routinely after every meal at a flower, passing their antennae through the cleaners several times before flying off to the next flower. Next time you're in a garden, watch them closely as they wipe their antennae clean right before they take to the air.

Antennae are vitally important sensory organs, which need to be kept clean. Though the forager bees spend hours every day outside in the sun, once they enter the hive, it is dark, and it is here that they use their antennae more than their compound eyes and three small, simple eyes, the ocelli, to sense their surroundings. Antennae bases are set in small socket-like areas on the bee's head, enabling them to move freely. Each antenna is connected to the brain by a large nerve that transfers crucial environmental information—touch, smell, taste, sound, humidity, and carbon dioxide levels. The antennae are literally covered with thousands of sensory organs receiving information. The mechanoreceptors are specialized for touch and hearing, the odor receptors for smell, and the gustatory receptors for taste. They "hear" by sensing the movement of air particles through the hairy mech-anoreceptors on their antennae. They listen closely to one another buzz during the waggle dance.

Bees move the pollen from the flower's anthers to their bodies in several ways. The bee uses its body hairs and legs to gather the pollen. As it moves over and through the flowers, the pollen is attracted to the bees' hairs by static electricity and sticks to them. It also actively scrabbles through the anthers to shake them and loosen the pollen grains.

The two forelegs are covered in hairy brushes used to wipe pollen from the bee's head. The middle legs transfer the pollen from the front legs to the hind legs, which are designed for the purpose of transporting pollen and tree resin. These are the legs that have corbiculae (New Latin: *corbis* meaning basket), commonly known as "pollen baskets," which are rather large concave areas with central bristles and a fringe of long, curved hair that can hold pollen grains or resin. Their forelegs brush pollen from their proboscis, sticky with nectar and regurgitated honey.

The honeybee takes to the air and, hovering over the flower, transfers the pollen from its forelegs to its rear legs, where it is passed over by pollen combs. The combs also remove pollen attached to the bee's abdomen. The pollen is then collected and transferred from the pollen combs to the pollen baskets by raking and scraping the inner surface of each hind leg. The bees press the pollen further by quickly rubbing their hind legs together in flight. Pollen moistened by the proboscis' regurgitated honey becomes stickier and thicker, turning it into a pollen pellet as more and more is added to the pollen basket. Next time you are near flowers being harvested by bees, look at the large green, yellow, orange, red, white, or blue (several flowers do produce blue pollen) pellets in their pollen baskets and watch them fly away, carrying their major source of protein back to the hive, where they kick it off into empty cells. The pollen-packing pantry workers will take over and press it into storage cells using their heads, mandibles, and forelegs, and when the cells are full, they will cap it with a thin layer of honey.

Pollen collection is regulated by the larvae, the nurse bees, and the foragers. Hungry larvae release pheromones that motivate the nurse bees to collect pollen from the storage cells. The foragers returning with pollen detect the emptying cells and communicate the need to continue pollen harvesting to other workers by performing vigorous waggle dances and making pollen-foraging trips themselves. If the pollen storage cells are full, these foragers cease their pollen-foraging trips and will only leave the hive to collect nectar, water, or propolis.

Bees are good listeners, using their entire body to hear what their nestmates are saying. They have organs in their legs that can detect vibrations passing through the comb. And sensors at the bases of their antennae are exquisitely sensitive detectors of airborne sounds (vibrations) in the hive. These vibrations are messages from other bees, letting the individual bee know what needs to be done to keep the colony alive.

In addition to his work on the dance of the honeybee, von Frisch also conducted a set of elegant experiments that revealed that bees see in color, although their eyes detect a different part of the spectrum than humans. They cannot detect red but are sensitive to blue, violet, and purple flowers and are exquisitely in tune with the ultraviolet spectrum, which we cannot see. Honeybees can use patterns of ultraviolet light to determine the direction of the sun even when it is hidden behind clouds. This is important when they are interpreting the waggle dance of returning foragers to determine the flower's direction relative to the sun on cloudy days.

Honeybees have five eyes: two large compound eyes and three small ocelli. The compound eyes are made up of thousands of single eye units. Each eye unit transmits its particular view of the world to the bee's brain, which translates the thousands of images (pixels) into a highly detailed portrait of the subject. The tiny hairs found between eye units help the bee determine wind direction and flight speed. Compound eyes lock on to anything that moves, and detecting movement can cause the release of alarm pheromones.

Each of a honeybee's ocelli has a single lens, similar to the human eye. The ocelli monitor the horizon and are used to maintain level flight. Through all these sensory mechanisms, the worker bees are able to find flowers and then efficiently harvest enough nectar and pollen to sustain the colony.

A worker honeybee collects nectar in a special organ called the honey stomach, or crop. As she sucks the nectar from the flower through her proboscis it travels through her esophagus and is stored in her honey stomach ready to be transferred to the honey-making bees in the hive. If she is hungry, she opens a valve in her honey stomach, and some of it will pass through to her "true" stomach, where it will be digested and metabolized into energy to fly back home to the hive. Her honey stomach can carry almost 70 milligrams of nectar, and it takes a lot of energy to carry an amount of nectar equal to her own weight, which foragers have to do if their honey stomach is full. Normally, though, a bee returning to the hive carries a partial load, a more efficient use of her energy, which she must care for in order to make the dozens of foraging trips during her short lifetime. To carry enough nectar back to the hive to be transformed into a 16-ounce jar of honey sold in the supermarket, tens of thousands of bees must forage more than 100,000 miles and collect nectar from 5 million to 10 million flowers.

Nectar is made mostly of sugar but contains some minerals, lipids, proteins, and vitamins, including a relatively large amount of vitamin C. Nectar can be fed directly to larvae, but it usually is processed into honey first. Harvesting bees carry the nectar back to the hive in their honey stomach and transfer it through regurgitation to the house bees, which hold the nectar on their tongues until the water evaporates. They then regurgitate the evaporated nectar into storage cells and fan it until it evaporates even more. When the amount of water in the nectar is less than 18 percent, it is protected from yeast. At the end of this process, the nectar has become honey and is capped and sealed by beeswax produced by the worker bees until it is needed to feed the hungry larvae.

Honeybee foraging with full pollen basket. *Leslie Day*

15

ACROSS THE STREET, BUT WORLDS APART

FEEDING THE HUNGRY AT ST. BART'S

> Privilege is the measure of duty, that the strong should help the weak.
> —The Reverend David Hummel Greer,
> St. Bartholomew's Episcopal Church, 1891

St. Bartholomew's reputation for helping poor and immigrant communities extends back to 1888 and has continued into modern times. In 2010, St. Bart's formed Crossroads Community Services to further broaden community involvement. Since then, Crossroads has managed St. Bart's soup kitchen, food pantry, and women's shelter. Each day a nutritious meal is served to 150–200 people, and 250 people take food from the food pantry each week.

Crossroads is run by the Reverend Edward Sunderland, LCSW, associate rector of St. Bart's. Father Sunderland is both a priest and a social worker, who, throughout his career, has specialized in medical social work, palliative care, and homelessness. Peter Betz initially reached out to Father Sunderland to explore ways the hotel could help, and when David arrived he developed the partnership. In explaining the Waldorf's history with St. Bart's, David explained: "One of the things I thought we could easily do would be to give them all the extra Danish, croissants, and bagels so they could use them for their breakfast. One of the cooks said, 'Well, why don't we just put them all in the boxes, like we do anyway at the end of the day, and just leave them on the counter in the kitchen. One of the people from across the street can just come over and get them?'"

David worked out the details with Waldorf security, telling them, "Hey, there are three guys that work across the street. If we give you their names is it okay if they come over every day?" It was that simple, David remembered: "They said, 'Yeah, no problem, as long as we know who they are.'"

David paused for a second and then continued:

> Here are these two buildings across the street from each other: the Waldorf Astoria and St. Bart's Church. They've been across the street from each other for more than eight decades, and over there you have hungry people and people that are trying to feed them. Right across the street is the back door to Oscar's restaurant. Now, instead of throwing prepared, untouched food into the garbage, we put it in a box. They walk across the street to get it every day. Baked goods, bacon, and sausage that have been on the buffet. Just yesterday, with the snowstorm and fewer customers, we had a lot of food and desserts left over from Oscar's. St. Bart's had about 150 people to feed.

Guests at St. Bart's soup kitchen. *Jeremy Daniel*

I never really thought how chefs feel about leftover food in their restaurants. David opened my eyes to the cook's feelings. "Chefs hate wasted food. If you think of all the people that are artists or create something, their work lasts. We're the ones that create something, and it's gone, it disappears, and too often it ends up in the garbage and chefs hate that."

The Waldorf regularly feeds ten women who sleep at the church's homeless shelter. It's a responsibility they share with the Lotte New York Palace Hotel and the Double Tree Hotel. Crossroads now has a new program born out of the collaboration with the Waldorf called the Crossroads Food Rescue program, which repurposes food from restaurants and provides it to soup kitchens and shelters.

In addition to giving unserved food from their restaurants to St. Bart's, the chefs also harvest fresh fruit and vegetables from the Waldorf's rooftop garden and share it with Crossroads' soup kitchen and food pantry. Added to the fruit are regular supplies of honey, harvested from the Waldorf's six hives.

16

APPLES, LAVENDER, TOMATOES, AND TARRAGON—EXPLORING THE WALDORF'S GARDEN WITH CHEF BETZ

> Flowers always make people better, happier, and more helpful;
> they are sunshine, food and medicine for the soul.
> —Luther Burbank

By early June, temperatures in the city are in the seventies during the day, dropping to the sixties at night. It was a good time to visit the garden and learn about the plants and how they are used in the chefs' dishes. When I arrived, the bees were out and foraging for nectar and pollen from the herbs, fruit, and the edible flowers of the garden—pansies, carnations, nasturtiums, lavender, thyme, dill, and cilantro. My nephew, Peter Betz, guided me from plant to plant, describing how he and the other chefs use the plants in their recipes.

The Waldorf's garden was lush with fruit trees and both perennial and annual plants, all of them feeding and being pollinated by honeybees from the six hives. Once the plants bloom, some bees will save energy by flying just a few feet to forage. Others will travel to ornamental fruit trees flowering on the streets below. Callery pear, crab apples, and cherry tree blossoms cover the city in April and May. Work begins in earnest in the hive by early April, as honeybees collect nectar and pollen to resupply the depleted hive.

Six apple trees grow in large containers on the Waldorf's roof, each a different variety. The pale pink blossoms do not produce much nectar per flower, but there are so many flowers that the foragers are able to reap reasonable rewards.

Apple trees are in the Rosaceae family, a group of plants that includes roses, pears, strawberries, almonds, apricots, cherries, nectarines, peaches, plums, prunes, blackberries, raspberries, and hawthorn trees. The six apple varieties are all variations of a single species.

Arkansas Black apples, so called because they are dark purple, taste crisp with a sharp flavor that sweetens with age.

September Wonder Fuji is a variety that shows a speckled pinkish-orange-red over a yellow-green background. They are juicy and so sweet they are considered dessert apples.

Winesap apples are small to medium with a deep cherry-red skin and crisp, yellow flesh. They can be eaten fresh or used to make cider, apple butter, and apple pies.

Ultramac apple is a Macintosh on a tiny tree that produces medium-sized tart apples.

Jon-A-Red is small and very red. The red skin "bleeds" into the white flesh. They are juicy, both sweet and tart at the same time, and used to make applesauce.

Rooftop apple blossoms with General Electric Building in background. *Leslie Day*

Lodi apples have yellow-green skin, a sweet-tart taste, and are used in both pies and applesauce.

Peter tells me that the Waldorf uses most of the apples in salads, including the famous Waldorf salad. They are also used to make cobblers by the pastry chefs.

A single Montmorency cherry tree grows in a large container on the roof. Cherry blossoms are the honeybees' best friends. Each flower produces an ample supply of nectar in one day. This type of cherry tree has ancient roots. Roman soldiers discovered it growing along the Black Sea in Asia Minor and carried the sour cherries with them throughout the Roman territory. As they moved, they planted cherry trees along the Roman roads. Cherries harvested on the roof are used in cherry pies, jams, and preserves and tossed into salads or become glazes for chicken dishes.

Several young fig trees also grow in the garden. Fig trees are one of the few plants on the rooftop that are not pollinated by honeybees. The only pollinator for the fig trees is a tiny, 2-millimeter-long fig wasp. In the wild, there are hundreds of species of fig trees, and each is pollinated by a particular species of fig wasp. The fig trees on the roof are *Ficus carica*, which is pollinated by the *Blastophaga psenes* wasp. This is a dramatic example of the evolutionary relationship known as mutualism, where each species benefits from the other, the same relationship that exists between bee and flower. The fig and fig wasp are more interconnected because they literally cannot reproduce without each other. This is called obligate mutualism, where the two species are "obligated" to rely on each other. The fig tree cannot reproduce without its single species of fig wasp, and the fig wasp cannot reproduce without this sole species of fig tree.

Among the many berries in the garden, the chefs chose to plant the *Vaccinium corymbosum* blueberry. Its common name is the sweetheart highbush blueberry, and it was selected because it produces two crops, one in early summer and one in autumn. They have been described as "fat, juicy blueberries bursting with fabulously sweet flavor."

The blueberry plants produce thousands of flower buds every year, and each bud can produce up to sixteen flowers. Every flower can potentially produce a blueberry. For this to happen, pollen grains produced by the anthers must be carried to the stigma so they can fertilize the ovules inside the ovary. Thus, every ovary with fertilized ovules becomes a blueberry.

The chefs grow six varieties of heirloom tomatoes in the garden. An heirloom is a plant that has a history of being passed down within a family or community, similar to the sharing of heirloom jewelry or furniture. Plants are verified as heirlooms by documenting the generational history of preserving and passing on the seed. As Peter and I lingered by the bed of tomato plants laden with small, sweet tomatoes that I kept tossing into my mouth, Peter talked about his Italian American grandmother's garden and how he loved to cook with her. It was part of his family's history that I was learning for the first time.

"When I was little everything revolved around cooking for my family on the weekends," Peter explained. "Every Sunday the whole family would get together at my grandmother's house. It was like a holiday. She would make octopus salad and zucchini fritters and all kinds of pasta dishes, braised meat, salmon, and all kinds of unusual meals, well not unusual in Italy, but unusual in America. She grew up in an Italian family in Brooklyn, and my grandfather grew up in Ischia, an island in the Gulf of Naples."

On those long-ago Sundays, Peter would be attached to his beloved grandmother's hip, watching her cook. He remembers that it was always a big production and that the end result was delicious. She never formally taught Peter to cook, but he showed curiosity and asked if he could help. He had an interest in preparing food as a child because of his love for his grandmother. So much of our connection to food seems to be tied to our connection to family, to those in our lives who showed a passion for greatness regarding the preparation of a single meal.

The garden's black plum tomato is a Russian heirloom that's brownish gray and tastes sweet and meaty. Peter tells me that the chefs often use them for making sauces, though they are delicious raw in salads. When mature, the oval fruits turn a deep mahogany-brown color.

The sugar lump tomato is an heirloom variety that dates to the 1800s and is considered one of the sweetest, small-fruited varieties ever developed. Originally from the Bavarian region of Germany, the long, grapelike bunches of sweet, red fruits measure only about three-fourths of an inch across.

The yellow plum tomato is a very old variety developed in the United States. The plants are large and open with small oval fruit, 1 by 1¼ inches. They have a mild but sweet taste. There are typically eight to ten fruits per cluster, and some late-season fruit have a slight neck so are considered plum shaped. The tomatoes are a lovely lemon-yellow color. They are excellent in salads. The plants produce fruit until the first frost.

The garden's Russian plum tomatoes are a Ukrainian heirloom, introduced to the United States in 1980. They produce an egg-shaped fruit with a smooth and perfectly unblemished skin that turns a deep purple-black in the hottest days of summer. Peter says they are one of the best varieties for salsa, as they have a "deep, rich smoky sweetness and a tangy aftertaste."

Green grape tomatoes were bred to have a rich, sweet, and zingy flavor by Tom Wagner, a renowned American plant breeder. The tomatoes are lime green inside with chartreuse-yellow skins. They are about the size of a large grape and are used for both salads and snacking.

The sixth variety is the orange oxheart tomato, a family heirloom tomato that originated in Virginia. It produces large, heart-shaped fruit that weigh nearly a pound. Possessing a brilliant orange-yellow skin, these tomatoes have a strong, aromatic "real" tomato smell. The Waldorf's chefs use these tomatoes for sauces and cooking.

The twentieth-floor roof has gone from being an area with air-conditioning units and peeling tar paper, to a magnificent, thriving chefs' garden—an ideal place for growing herbs because it basks in sunshine throughout the day. With the final addition of more garden beds, the garden now has twenty-three, waist-high beds buzzing with bees, emitting scents of lavender, mint, sage, and other herbal fragrances. The hotel now hosts events in the garden, including the annual Battle of the Bees, where guests taste honey from local New York City apiaries and, in a blind test, choose their preferred honey. During the first Battle of the Bees I attended in 2015, the Waldorf was the winner. In 2016, the Whitney Museum's apiary was selected as having the best honey.

The smell of mint and chocolate attracted me to the next bed where Peter showed me *Mentha piperita* growing. This is a chocolate mint whose flowers are flowing with abundant nectar loved by the honeybees. The nectar of this perennial gives honey a mild minty flavor. Its pale, extremely fragrant purple flowers open from the bottom and spike up. Chocolate mint is commonly grown as a culinary herb and can reach two feet tall. At the Waldorf, this plant is an ingredient in their mint chocolate chip cookies.

Lavender is another member of the mint family. It is an evergreen, maintaining pale green leaves throughout winter. Its strong essential oils repel pests, so it serves as a hardy companion plant, keeping munching insects away. The chefs use it to make a honey glaze for the chicken and as an ingredient for homemade lavender ice cream.

The garden's lemongrass is native to the Philippines and Indonesia where it is known as *tanglad*, or *sereh*. Its leaves are extremely fragrant and contribute a mild sweetness and lemony flavor, whether in chicken dishes or tea. It is a perennial and evergreen grass often used as a lemony flavoring in recipes from Thailand, Vietnam, Laos, and Cambodia. It is also an ornamental grass, with gracefully arching three-foot-long, pale-green leaves. The lemony scent is released when you gently bruise or press the leaves. It's of little use to the bees, as it does not often produce flowers, but luckily for the restaurant and hotel guests, it is used frequently as a fresh ingredient for Asian dishes. Lemongrass is sometimes used in perfumes and in herbal medicine. Peter told me that the Waldorf chefs "use lemongrass a million different ways. We use it in a coconut curry sauce for a lobster dish that we make with rice wine, coconut milk, red curry, ginger, and lemongrass."

Levisticum officinale, commonly called lovage, is a large, beautiful plant with shiny, finely cut compound leaves that look and smell like celery. Lovage blooms in the rooftop garden during early summer, producing umbels of yellow flowers, which are extremely attractive to the honeybees. Eastern black swallowtail butterflies lay their eggs on the leaves, as it is their host plant. Their caterpillars nibble the leaves, but do not defoliate the plant. After pupating in their chrysalis for a week, they emerge as large and gorgeous black butterflies with yellow spots and powdery blue hind wings.

Every part of lovage is edible for humans. The large stems are often candied, and the young shoots are used in salads. Its seeds are added to pastry and sprinkled over fruit. The root and other parts of the plant are often used as herbal remedies.

The Italian oregano grown in the garden is also known as hardy marjoram. It's a cross between sweet marjoram and wild marjoram, bringing together flavor and hardiness. It is a bushy evergreen perennial with dark-green, aromatic leaves. The plant can grow to two feet and flowers throughout the late spring and into early autumn. The pinkish white flowers attract many pollinators, honeybees among them. The flowers are hermaphrodites, having male and female parts. The seeds ripen from August through September.

The leaves from the Italian oregano can be eaten fresh or cooked. They are used to flavor salad dressings, vegetables, and oils. Its highly scented seeds are used to flavor sweets and drinks. They taste like a lovely blend of thyme, rosemary, and sage. The plant's oil of sweet marjoram is used in food flavoring and as an aromatic in perfume, soap, and hair products.

Oregano is native to the areas near the Mediterranean Sea. It is a plant that loves the sun, and its flavor only grows stronger under the hot summer sun on the roof garden. The chefs of the Waldorf often use this herb in their tomato sauces.

In Greek mythology, the goddess Aphrodite brought oregano to mortals to make them happier. In translation from Greek, *Oros ganos* means "joy of the mountains." Newlyweds in ancient Greece wore tiaras woven from oregano leaves. Thinking that the plant would pave the way to the next life, the ancients planted oregano around graves. It was also said that the herb had magical powers that would protect against evil and bring joy to life, so people would put it in their beds.

Romans carried oregano with them throughout Europe to use when cooking fish and pork and to flavor their wine. Americans discovered oregano when returning soldiers brought it back from Europe after World War II. In part due to the popularity of pizza, America's love of oregano grew to what is today.

One of Chef Garcelon's favorites in the garden is French tarragon. Its leaves have a strong, anise-like flavor and aroma. It is a shrubby perennial, which can grow tall and reproduces vegetatively via creeping rhizomes. It rarely flowers. Tarragon's narrow, lance-shaped, dark-green leaves may be used fresh or dried to flavor fish, meat, vegetables, eggs, salads, sauces, or vinegars. It is the primary flavoring in sauce. French tarragon plants rarely produce seeds, but when they do, the seeds are sterile. Plants are propagated by rooting stem cuttings or dividing rhizomes. The tarragon leaves are much loved and valued by the chefs at the Waldorf for their incredible fresh taste and aroma. Peter explained that "tarragon is delicious with Peruvian potatoes. We toss the potatoes in good Spanish olive oil, a little bit of sherry vinegar and tarragon. The tarragon is so flavorful, so aromatic, you don't need much."

I asked Peter how he went from loving cooking in his grandmother's kitchen to becoming a chef, a part of our family history with which I was only vaguely familiar. Like David Garcelon, Peter took a part-time job in a restaurant to earn extra money when he was a student. He loved it. From there, he enrolled in a continuing education program in Storrs, Connecticut.

His classes were geared toward professional cooking, which he said is very different from cooking for your family.

"You have to love food and you have to know how to put flavors together," he said. "The actual processes of cooking in a professional kitchen are different: the volume and timing and working with so many different ingredients at one time. I really liked it." The school opened the kitchen to the public to make money for the program and to give the student chefs a professional experience. Senior citizens would come for lunch and pay $5.00 to eat in this mock restaurant. It was here that he learned line cooking. "You have a physical line in the kitchen," Peter said, "and on one side you have the cold station, salads and sandwiches, and on the other side, you have the hot station: grilled and sautéed foods. Classical French cooking is a little different, but in typical American restaurants, you have stations, and each chef is responsible for their station and a certain number of menu items." Some chefs like working in a certain station and stay there their entire career, but Peter wanted to learn all of it—every station. "I learned every discipline. I learned the role of the saucier: making sauces. I learned the sauté station, and the grill station."

A friend encouraged Peter to think about going to a serious culinary school and told him about the renowned Culinary Institute of America in Hyde Park, New York. He visited the institute and was taken with the program and its beautiful rural setting, 170 acres of undulating hills along the eastern banks of the Hudson River. The school first opened its doors in New Haven, Connecticut, in 1946 after World War II, calling itself the New Haven Restaurant Institute. It was a vocational school for returning war veterans, offering a sixteen-week program. Eventually, as the restaurant industry grew, its enrollment swelled. The school found a new home in a former Jesuit novitiate, St. Andrew-on-Hudson, in Hyde Park, New York. Today, the institute is the only school authorized by the American Culinary Federation's Master Chef Certification exam. The faculty, including 125 master chef instructors, hailing from sixteen countries, teaches almost 3,000 students. The campus sits near the Hyde Park home of Franklin D. Roosevelt, which is a National Historic Site run by the National Park Service and given by Roosevelt to the federal government as a gift to the American people in 1945.

Peter applied to the Culinary Institute of America and that summer worked in the North Street Grill restaurant in Great Neck with Chef Brendan Walsh, the restaurant's co-owner and a graduate of the institute. Chef Walsh had some fame as the former executive chef at Arizona 206, the Manhattan restaurant. Walsh was the first legitimate chef with whom Peter had worked. In 1992, Peter started culinary school and lived in the dorm.

After the first year, he embarked on a six-month externship at the Garden City Hotel's restaurant in Long Island. He then completed the second year at the school's restaurant row, working in all of the restaurants. St. Andrews was named for the former Jesuit novitiate and was a café with healthy food, so Peter learned to substitute rich ingredients with lighter and healthier items. At Ristorante Caterina de' Medici, he learned Italian cuisine before moving to the famous and now closed Escoffier Room, where he learned French cuisine. His final restaurant row experience was at the American Bounty.

Peter described his experiences this way: "During that process in restaurant row, you are working in the kitchen and at the front of the house. You are learning everything about the restaurant business. These are famous restaurants in their own right. You have to complete your tour of duty in each one of these restaurants in order to graduate." And graduate he did.

In 1994, Peter took his résumé to the Four Seasons Hotel on East 57th Street in Manhattan, which had just opened. He loved working in the Fifty-Seven Fifty-Seven Restaurant at the Four Seasons, which, two years later, earned three stars from Ruth Reichl at the *New York Times*, which, he said "is no easy feat. I think at that time there were fewer than twenty restaurants in New York City that had three stars, and we were a new team in a new restaurant." Peter said he was greatly influenced by Susan Weaver, the executive chef at the Fifty-Seven Fifty-Seven.

After a few years, Peter moved to the Waldorf Astoria, when the hotel reopened the Peacock Alley Restaurant with the celebrated French chef Laurent Gras. "He was my draw to come here," Peter explained. "Everyone wanted to be on the property just to be around him and see what he was doing. He was a supertalented guy." Laurent Gras and Peacock Alley also received three stars from *New York Times* reviewer Ruth Reichl.

From Peacock Alley, Peter went on to work in different restaurants at the Waldorf Astoria. They had just renovated Oscars and asked him to work there and help reopen it. He loved working with Executive Chef Joe Verde and decided to stay at Oscars. Until then Peter's title had been "cook," but once he started working with Chef Verde, because of his degree from the

Culinary Institute and his experience at the Four Seasons restaurant, they gave him the title of sous chef, which is a supervisory position. *Sous* means "under"—directly under the head chef. Peter said, "The chef is the visionary, the one who comes up with the dishes, comes up with the concepts, comes up with the directions and then the sous chef carries out these operations, seeing the chef's vision to fruition."

I asked Peter about these French terms, and he told me they are part of the French brigade system codified by August Escoffier to simplify and organize the roles of a kitchen's staff from top to bottom: from the executive chef to the dishwasher.

From Oscar's, Peter moved up to the banquet kitchens. The banquet operation at the Waldorf is massive, with six venues serving thousands of guests. There are several sous chefs directing the cooks and waitstaff to serve banquet guests. On any one night, they may serve food to 1,500 people in the Grand Ballroom, 260 in the John Jacob Astor Salon, 260 in the Jade Room, 120 in the Basildon Room, 400 in the Empire Room, 380 in the Vanderbilt Room, 600 in the Starlight Roof.

From there, Peter was promoted to the chef garde manger, responsible for the cold station for banquets, which includes preparing all of the hors d'oeuvres, appetizers, canapes, pâtes, salads, and terrines. *Garde manger* is French for "keeper of the food," and the cold station banquet foods are prepared in a well-ventilated area and kept on the cool side to maintain freshness.

Two years later, he was promoted to banquet chef, in charge of the hot station, the cold station, the entire preparation for every banquet, with ten sous chefs under him. Peter started developing strong relationships with the clients who, year in and year out, booked their banquets with the Waldorf. The Archdiocese of New York, the March of Dimes, Bette Midler's New York Restoration Project, and others all depended on Peter to set up and run their annual fund-raising events at the Waldorf. So when the hotel needed an executive sous chef, the chef directly under the executive chef, they offered the job to Peter, but they also asked him to keep his job as executive banquet chef. He ended up wearing both hats. Then the Waldorf hired David Garcelon to be executive chef and director of culinary. Peter became David's executive sous chef and continued as the head banquet chef.

Peter and David were a good fit because their personalities complemented each other and they worked together with a natural ease. They are both intelligent, hardworking, creative, and caring men who support each other and their team. Whenever I walked through the halls and rooms of the Waldorf with Peter, everyone greeted him with the same warmth that David received.

> When David got here, we worked really well together, even though there was a lot of work to do. But the ease with which we collaborated freed us up to think more creatively. Before he arrived, the chefs were so buried in the operation that it was like we had tunnel vision in order to do our jobs well. It was hard to see past tomorrow. David brought a new sensibility to the hotel. No matter how busy we were, we had to think about the future. I really admire that about him. He had experience with honey-bees and a chefs' garden in Toronto. Once he decided to go ahead with the Waldorf garden, he conceptualized it for us, and we all got to work on it. Everybody had a part to play, but David was the leader. We were all so excited about it. Everyone was willing to chip in whatever it took to get it done. People participated to varying degrees, but I would say that everybody participated at some level. Some were involved in building and painting the raised wooden garden beds. We didn't know anything about honeybees, or building their hives, but we all wanted to learn.

They learned how to construct a hive, harvest honey, and everything else that goes along with beekeeping. They cared for the bees from the beginning of the season in the spring to the end of the season in the fall. Peter said, "I don't consider myself an expert, but I am excited by all I've learned from our expert, Andrew Coté. And the end result is that we have this fresh, terrific honey that we can use in our recipes. This has all added a level of excitement to the crew. It was motivational."

In creating recipes, Peter said that he uses the rooftop honey to enhance food. He talked about the effect honey has on food and how it complements a dish: "One of the first foods we realized that honey changed for the better was ice cream. There are certain foods that add an additional note, an additional depth." I asked him what he meant by note.

"A note in cooking means an additional dimension," he replied. "If you have five or six flavors that come together, each one will be a different note, like in a song, and in this case honey adds one note to it, but all the notes together form a new flavor, a new song. In the case of ice cream, the honey flavor comes out so naturally that it enhances the texture and temperature of the ice cream. In some dishes, we look for that and in some dishes we want the honey to be subtle. We use it to add to the taste, but it is not the main feature."

Peter had never looked at honey or used honey that way before. During his education at the Culinary Institute of America twenty-five years ago, he said that they did not keep honeybees.

"The buzzword back then was 'imported.' Now it's the other way around," he said. "The buzzword today is 'local.'" Peter thinks that it is cyclical. "If you grow up around all these local ingredients, at some point you want to try something from overseas. And, eventually, you want the best of both worlds." This is another way of saying what David Garcelon said: food is complicated.

Peter added,

> That's what cooking is now. It has so many different influences. There are restaurants dedicated to different regions of the world, to be sure. But more often than not these days, especially in the United States, restaurants are using a mix of ingredients: Asian, Indian . . . it's fusion, using ingredients from different regions: maybe calamari with sweet chili sauce. In my kitchen, we have people from every part of the world. We have 140 cooks plus we are influenced by people who come by my department and say. "Oh Peter, I want you to taste something I made in my house last night." In our events, we always have a starch and a vegetable with whatever protein item we serve. In our banquets we are offering an appetizer, an entrée, and a dessert. Guests have a choice of different items, and we try to make it as well rounded as possible. We are always looking for dishes that work well together and are new and interesting. One of our employees was from Thailand and brought in coconut sticky rice balls infused with lemongrass and ginger that he had made at home. We used his recipe and made a galette out of it.

I asked him to define *galette*, and he said it's a French term for a pancake-like, sweet dish. It refers to its shape: flat and round. He went on, "You wouldn't think of rice in that way, so we adapted his Thai dish from a sticky coconut rice ball into a pancake with those same flavors. He brought in this dish from home and we took his idea and adapted it into something that was totally different."

In Peter's kitchen up to thirty countries are represented at any given time. Peter said, "The diversity is amazing. And so many of our staff are interested in the bees and the garden. Some go up every day and pick herbs and incorporate them into a recipe they are preparing. Others might go up every day just to collect flowers or leaves or sprigs to decorate platters for our receptions. People participate in the garden because they are so interested in it."

The garden evolved over time. The chefs added planters and plants that they believed would enhance their cooking. As we walked around the garden on that June day, Peter stopped by a bed with a lovely and fragrant species of sage. It was variegated sage, also known as tricolor sage. It adds beauty to the garden with its green-, white-, and purple-tinged leaves. Sage has been an important part of European kitchen gardens for hundreds of years. A native of the Mediterranean, and cultivated for culinary and medicinal use, it is a very hardy plant. It has been written about as a culinary herb since 1597. All parts of the plant have a strong scent due to the essential oils that carry the chemical essence of the plant's fragrance. In the Waldorf's kitchens, sage is often used as a garnish. In the autumn, the chefs like to use it in stuffing. From late September and into October, the chefs prepare chestnut sage stuffing with croissants chopped into it.

Sage gets its name from the Latin word *salvere*, which means "to save," a reference to its medicinal properties. Ancient Greeks and Romans considered it a sacred herb, and it was used to prevent food from spoiling. Sage grows to be about a foot tall with finely wrinkled leaves arranged in opposite pairs. The leaves are hairy with glands that store the volatile oils. Purple flowers are arranged in whorls and bloom in August. Bees devour the nectar and produce a sage honey that is highly valued.

Peter pointed out the garden thyme, *Thymus vulgaris*, a woody perennial that bears tiny, pointed, gray-green leaves, with their edges rolled under. They form a mound six to twelve inches tall. Just before flowering, the leaves become highly aromatic and are used fresh or dried to season soups, stews, sauces, meat, and fish. In late spring to early summer, delicate purple flowers emerge from the ends of the stems and are visited by the honeybees.

Chef examining a strawberry, Waldorf Astoria roof garden. *Leslie Day*

Pliny the Elder, the Roman author, philosopher, and naturalist, wrote about thyme honey in his *Naturalis Historia*, published AD 77–79: "the honei which commeth of Thyme, is held to bee the best and most profitable: in color like gold, in taste right pleasant."

Thyme has another use, one unrelated to cooking, but of importance to beekeepers. Varroa mites are parasitic invertebrates that live on honeybees. They puncture the bee's exoskeleton and suck its blood, called hemolymph. The active ingredient in thyme's essential oil is thymol. Beekeepers use it to help destroy varroa mites. It works by confusing the mite and blocking its pores. When using thymol, beekeepers install a screened bottom board, and when the mites become confused, they fall to the ground through the screen and cannot climb back up into the hive.

If beekeepers are unable to control varroa mites in their hives, the wounds caused by them expose the bees to a number of viruses, some of which are suspected of causing colony collapse disorder, when the majority of a hive's worker bees suddenly disappear. This syndrome has devastated millions of hives in the United States, particularly since 2006. Although the causes are still being investigated, some believe that bees are more vulnerable to mites because of their exposure to neonicotinoid insecticides, which are used as pesticides on agricultural crops.

Looking into a different bed I notice a familiar little plant that grows throughout the parks, apartment grounds, and backyards in New York City: *Oxalis stricta*, common yellow wood sorrel. It is a low-growing plant that looks like shamrock, with a compound leaf made of three tiny leaflets. It bears small, yellow flowers. It is used as a flavoring for a fish nage, a broth that sometimes accompanies fish served at the Waldorf. Many gardeners treat wood sorrel like a weed and pull it out, because it grows wild and is not typically planted. But it has a delicious lemony flavor and those in the know are not surprised to learn the chefs planted it on purpose.

As we finished our tour of the garden, I took one last look around. A lot of effort had gone into building it; that was clear. But there was something more here. Life. They had brought life to a place that was previously lifeless. These healthy bees flew from flower to flower, the aroma was rich, the colors of the flowers were beautiful, the feelings I had were soft and special and I felt content.

WINGS!

> That thousands of insects could live together so densely
> and harmoniously, and could build delicate wax combs filled
> with delicious honey, was almost a miraculous wonder that left a
> deep impression. No less impressive was what I saw when I lay in the
> tall grass beside these hives: thousands of humming bees
> crisscrossing the blue summer sky like shooting stars.
> — Thomas D. Seeley, *Honeybee Democracy*

During flight, the wings of worker bees beat more than 230 times per second, a seemingly impossible achievement. Their flight muscles contract so often every time their nerves fire, that they continually vibrate, causing the bees' internal temperature to rise to 36 degrees Celsius, or 97 degrees Fahrenheit. To avoid fatally overheating, the bees can regurgitate droplets of nectar onto their heads, which prevents overheating through evaporative cooling. As the water in the nectar evaporates, it carries the heat away from their bodies, just like our sweat does, thus cooling them off and allowing them to fly hundreds of miles in their short careers as forager bees collecting pollen, nectar, water, and propolis.

The honeybees' wings are outgrowths of their cuticle, or exoskeleton: an adaptation for flight. Each bee has two pairs of transparent wings attached to the thorax by means of highly complex joints, which allow them a great range of movement. The front wings are larger than the hind wings, and the bee can attach the wings together by coupling them with hooks called *hamuli* so that they beat synchronously, which greatly reduces turbulence and drag during flight.

These transparent, delicate appendages are strengthened by the veins running through them, which, when the wings were being formed during development, carried blood, oxygen, breathing tubes, and nerves to every part of the wing but in adult bees serve simply as support.

Wings are used for much more than flight. When thickening the nectar, the bees fan air through the hive with their wings. It takes countless wing beats to evaporate the water in nectar to make honey. Nectar contains 50 percent to 70 percent water, whereas honey contains less than 20 percent water. When many flowers are in bloom and the bees gather plentiful nectar, they fan all day and they fan all night, working in teams to concentrate the nectar and turn it into honey.

The air-conditioner bees help by keeping the air flowing throughout the hive so that the moisture removed from the nectar is blown out of the hive. When the job is done, these hardworking bees are rewarded with thick, sweet honey.

During the hot summer, fanning bees have their work cut out for them. On steamy summer days and nights, these hardworking bees line up at the hive entrance and fan their wings like crazy to move air around the hive. Incredibly, they bring droplets of water inside the hive opening, fan their wings over the water, and blow cool air throughout the hive. They are tiny air-conditioning units.

During summer, the bees must also keep their wax combs from melting. On very hot days, the returning foraging bees leave water droplets on the combs and fan their wings, cooling down the wax. The wings of honeybees

Female honeybee's worn-out wings. *Leslie Day*

wear out quickly because they beat hundreds of times every second when a bee is flying outside the hive or fanning inside it, making the air vibrate, which is the buzzing sound we hear.

There are times that the guard bees use their wings to defend the hive. When small ants try to invade the hive for its honey, the guards turn around in front of the ants and fan their wings so hard and fast that they literally blow the ants off their tiny feet and out of the hive.

On average, worker bees can fly about fifteen miles per hour. To have the energy needed for this Herculean task, the thoracic muscles must be supplied with enough energy from sugar in the honey they consume. Before they even take flight, foraging bees take a sip of honey and store it in their honey stomachs, metabolizing it on their way to and from the flowers they visit.

The thorax is the middle segment of the insect's body. It is where the six legs and two pair of wings are attached. It is chock full of muscles that are needed to power the wings. And the source of this power is the honey carried in the bee's honey stomach. This organ, also called the honey crop, is located in the abdomen. The long, narrow esophagus connects the crop to the bee's thorax and then to its proboscis. The honey stomach expands to carry a small amount of honey from the hive, which powers the bee's flight to its nectar source, and then it expands to carry nectar from the flower back to the hive to be turned into honey. The honey stomach is also used to carry water back to the hive. When the flight muscles need energy, the honey or nectar stored in the honey stomach supplies sugar directly to the flight muscles.

Bees move their wings in a very complex and precise manner. They can rotate their wings, change their angle, hover, move forward, move backward, and make subtle turn maneuvers. And they can do all of this at 230 beats per second.

During the flowering season when a worker bee reaches the age of three weeks or so, she starts to make trips outside the hive, and for the rest of her short life, her work will consist mostly of making flights to and from flowers, gathering pollen and nectar. On average, each bee will fly some 300 miles and will literally work themselves to death. Their wings—their entire bodies—become worn out from this labor. To keep their hive safe and clean, they will take their final flight and die away from the hive. Though hundreds of bees may die in one day, even more larvae hatch from their tiny eggs. And the life of the colony goes on.

FARE SHARE FRIDAY

NEW YORK CITY HOTELS GIVE BACK

> Fare Share Friday is a taste of a world without hunger and homelessness.
> —The Reverend Edward Sunderland, Executive Director, Crossroads
> Community Services at St. Bartholomew's Church, New York City

O n the Friday after Thanksgiving, beneath the massive nave of St. Bartholomew's Church, surrounded by treasures of sacred art, mosaic masterpieces, sculptured marble, lustrous wood, and stained glass windows, 500 people have gathered. Half of them are poor or homeless and half are paying guests. They will all be fed by the executive banquet chefs of five of New York's best hotels. David Garcelon and Peter Betz of the Waldorf Astoria Hotel, Jacques Sorci of the Lotte New York Palace Hotel, Willis Loughhead of the International Barclay Hotel, John Johnson of the Four Seasons Hotel, and Gene Harran of the New York Hilton Hotel.

This event was created by Rev. Edward Sunderland, executive director of Crossroads Community Services, the chefs, and the Sustainability Committee of the Waldorf Astoria. It is a spectacular social gathering called Fare Share Friday, where mouthwatering gourmet food is served with respect to some of the city's hungry and homeless people.

The idea of Fare Share Friday occurred to Rev. Sunderland in 2010 when he asked the people who use the food pantry and come for breakfast, those he calls the church's "guests," what they wanted on Thanksgiving. According to Father Sunderland "they pointed out that many people provide food on Thanksgiving, so they asked that we offer a Thanksgiving meal on the day after. Initially, volunteers served the meal in a conference room at St. Bart's. In 2012, David asked if the Waldorf could provide the food and of course I said yes."

By 2014, David helped recruit Jacques Sorci, executive chef of the Lotte New York Palace Hotel, to also donate food to St. Bart's charitable programs. That was the year that Sunderland shared his idea of enlarging Fare Share Friday by inviting paying guests. He talked it over with the chefs, the people who are fed by the church, and board members Carolyn Gargano and Laurel Dutcher. He asked them, "Why not invite people of means to Fare Share Friday to build relationships. The money would be used to pay for the event. Additional money raised would go for special programs."

In an interview with the Living Church, Sunderland said, "The Church can be a place where businesses can come together with people who are interested in solving the problems of hunger and homelessness, along with the people who are living the problems of hunger and homelessness."

Advertising agency Saatchi and Saatchi Wellness and their art director, Carolyn Gargano, also a board member of Crossroads, have been involved from the beginning and designed a graphic icon for this gathering. According to Carolyn, their goal was to promote "a special event to unite the community around the mission to end hunger on the streets of New York."

The logo they created, pro bono, is of five spoons filled with pistachio nuts, almonds, brown sugar, cranberries, and candied orange rinds. The spoons are surrounded by the treasures of the Waldorf Astoria's chefs' garden, all picked fresh that day for the photo shoot: heirloom cherry tomatoes, gooseberries, parsley, carrots, figs, sprigs of lavender and rosemary, thyme, purple basil, Swiss chard, oregano, lemon thyme, fresh dill, cherry pies bursting with cherries from the rooftop cherry trees, wild strawberries, and chunks of honey comb gleaming with golden honey from the hives of the Waldorf's honeybees.

The logo of the five spoons was something Carolyn created after everyone settled on the name, Fare Share Friday. Carolyn had been volunteering at the Women's Shelter at St. Bart's, and one of her jobs was to set the table for dinner.

> I noticed the importance of this simple act to welcome diverse women, each with a unique story. The spoon became a symbol because it not only represented eating, but it was for many of

us the first utensil you were introduced to as a child, so it has a nurturing, almost maternal quality. Its round shape is so different from the sharp fork and knife. I thought about all the diverse people Crossroads feeds—thus four of the spoons represent different economic classes—plastic, stainless, pewter, and silver. In the center . . . the chefs' spoon. The spoon inspired by the kindness of the chefs at the Waldorf Astoria who were the first to work with Crossroads to feed the homeless. I saw setting the tables in the shelter as an act of service.

The image of the five spoons filled with nuts and fruit was created in the kitchens at the Waldorf Astoria Hotel. Carolyn explained that some of the chefs and staff

> kindly rose to the challenge when I asked them to bring the logo to life. The honey is from their beehives, as well as all the ingredients and little pies from their kitchen. The image was created over the course of nine hours by a food photographer and his stylists, who sat helping me, and Charlie Romano, the Waldorf's executive pastry chef. Every almond was graded A to C for quality by hand. We only used A quality. Every nut, berry, and candied orange rind was placed with tweezers. The sugar cubes were carefully shaved at the edges by Charlie. The Waldorf made the stencil, and Charlie Romano painstakingly added wheat flour over each of the letters.

Others dove in to help create this magnificent evening. David's wife, Kylie Garcelon, a professor in the Hospitality Management Department at the New York City College of Technology, brought several student chefs with her to assist the hotel chefs with preparing and serving the meals during Fare Share Friday. Interviewed by CUNY (the City University of New York of which New York City College of Technology is a part) *Newswire*, Professor Garcelon said, "An important theme in the Hospitality Management Department is the spirit of volunteering—of giving back. In fact, in our department we have a strong culture of this, of which we are immensely proud. In hospitality education we are perfectly poised for this type of community involvement. The students gain exciting and valuable industry exposure while giving back to their community."

The men and women who volunteer in St. Bart's soup kitchen every day, some starting as early as 5 a.m. and working there for a few hours before they go to their own jobs, were also helping out.

Michael Romei, head concierge at the Waldorf Astoria, brought dozens of colleagues from hotels all over the city through the New York City Association of Hotel Concierges to help organize and run Fare Share Friday.

Laurel Dutcher, a Crossroads board member, whose career with the Hunger Project has been dedicated to ending world hunger, said of the night: "Like Amazing Grace, like Brigadoon, like entering the Land of Oz . . . it was like lifting the veil and suddenly seeing all of us together in a collective simple moment of clarity and peace — an affirmation that what we long for is already right here among us."

In the retail world, this day is known as Black Friday, where shoppers line up outside stores before dawn to grab sale items. But aside from sharing a square on the calendar, Black Friday and Fare Share Friday have nothing in common. The money raised by paying guests helps feed a highly nutritious breakfast to 150 people each morning of the year, to feed and shelter up to ten homeless women that St. Bart's cares for each night, and to feed almost a thousand people a month through their food pantry.

Tables set, waiting for guests to arrive. *Leslie Day*

The church is filled with twenty-five large, round tables for the event, each covered in a scarlet tablecloths and decorated with gorgeous holiday floral centerpieces. They will be changed, as there are two sittings: one at 4 p.m. and one at 7 p.m. All of the hundred-plus volunteers wear white shirts and black pants. I experienced the event firsthand, as my husband and I helped serve from 2014 through 2016. Gene Harran, the calm and efficient director of banquets from the Hilton, taught us how to lay out a proper place setting with two forks, two knives, two spoons, a water glass, a punch glass, a red napkin, several plates, serving utensils, and candles.

Once the guests arrived, we lined up and filed into a side area where the food was waiting and plates were handed to us by the chefs. Among our tasks, my husband and I carried out the seasonal salad created by Jacques Sorci, executive chef at the Lotte New York Palace. It was made with cranberries, chestnuts, field greens, and an apple cider vinaigrette. The salad was preceded by an appetizer and followed by turkey, gravy, potatoes, vegetables, cranberry sauce, stuffing, bread, and butter. The smell of the food, and the gorgeous presentation, punctuated by the care the chefs took in preparing the meal, was magical. During dinner, we filled water and punch glasses and served additional food whenever a guest requested it.

Long tables in the middle of the nave held decorated desserts that the diners could select after they finished their dinner. They were also given takeout containers to fill up with leftovers and desserts—as many as they wanted. Tonight, I recall thinking, everyone here has been enriched. Not just materially, but in the deeper meaning of the word.

During the 2016 serving, Jim and I decided to try something different. We helped serve during the first sitting, but we bought tickets for the second, in order to experience the meal from a diner's perspective and to connect to guests in a different way. As our table was seated, there were some quiet moments, typical when strangers gather. But once people started passing around the turkey and gravy, the awkwardness melted away and neighbors spoke to neighbors.

Soon, the large room was abuzz with conversation among the homeless, law firm partners, finance executives, teachers, writers, artists, musicians, retirees, and others from every profession and every walk of life. That night we experienced an evening of compassion, understanding, and plenitude. A woman sitting near me, who had just turned seventy, told me she had been a seamstress in the garment center most of her life, and now lived in an SRO

Kylie Garcelon setting up desserts. *Leslie Day*

(single room occupancy) hotel. She comes to the church's food pantry every morning to have her breakfast. Her eyes were sparkling as she looked around the glorious room and said, "This is a Cinderella night."

Calogero Romano, known as Charlie, the talented executive pastry chef at the Waldorf said, "What I love about baking for this event is that it's in the fall, with all the spices—like the scent of cinnamon—filling the kitchen with amazing aromas. Then there are all the other ingredients. To feed 500 guests for dinner, it requires at least 150 pounds of sugar, 200 pounds of butter, and 275 pounds of flour." What Chef Romano and his team do with those hundreds of pounds of butter, sugar, and flour can only be compared to what the great painters do with pigments, what the great sculptors do with stone and metal, and what the great composers do with notes. The artistry of these pies, cakes, cookies, and other sweets is as miraculous to see as they are to taste.

The dessert buffet is home to pumpkin and chocolate éclairs, pecan pie, pumpkin pie, apple pie, red velvet cake, cookies, and other iced and decorated sweets. The tiers of desserts are presented on gilded serving dishes. Executive Pastry Chef Jasmina Bojic from the Four Seasons Hotel made chocolate cake and a tiramisu with mascarpone cheese.

When the Waldorf chefs were planning the first Fare Share Friday, there was some discussion about whether they should even have a dessert buffet instead of a dessert service: would people behave or would there be pushing, shoving, and grabbing because they were homeless and hungry. David recalled that he "wasn't worried about that. I said, 'You know what, it's not going to be any better or worse than it is at the Waldorf Astoria,' where there are sometimes people who push and shove and grab." He went on, "In the end, oh man, everyone lined up waiting their turn, respectful and careful. It was great—you couldn't ask for it to be better. Somebody at the planning meeting had said, 'If you treat people with respect they will be respectful.' And that's really what it was like."

A lot of money is raised on Fare Share Friday. David explained it this way after the first event in 2014: "There are some really good things happening with Crossroads. The money we raised from Fare Share Friday is pretty significant, and there's a hope that we can do that again next year. We're talking about an organization with a $400,000 a year budget, so raising $80,000 in one night is a pretty good shot in the arm. And they are hoping to expand their programs to serve more homeless people." In 2016, the event exceeded expectations, and more than $100,000 was raised to help feed and shelter the homeless and poor at St. Bart's Church.

David told me that Jacques Sorci, the chef from the New York Palace, sent him an email after the most recent event. "Jacques is a French master chef who has been around a long time. He told me that it was one of the most rewarding meals he has ever served in his life."

Fare Share Friday menu.
Leslie Day

Before my participation in Fare Share Friday, my volunteer experiences had never included helping local shelters for the homeless. I learned that this particular way of serving is therapeutic, calming, and rewarding in ways that felt different from the educational and environmental volunteer work I do. Lining up and being told exactly what to do in service to others was both fulfilling and comforting. Being part of a small army of people engaged in labor to feed neighbors in need of not just food but of respect was deeply satisfying. A sense of peace descended upon me as I grabbed a pitcher filled with water and walked with it toward a table of people. I was happy to see an empty glass that needed filling, and I felt lifted every time the person I was serving said, "Thank you so much."

SAVING THE TREASURES AT THE WALDORF

> *You're the top!*
> *You're a Waldorf salad.*
> *You're the top!*
> *You're a Berlin ballad.*
> *You're the boats that glide*
> *On the sleepy Zuider Zee,*
> *You're an old Dutch master,*
> *You're Lady Astor,*
> *You're broccoli!*
> —Cole Porter, "You're the Top"

New York is a paradox, a city that loves to change and a city that loves to preserve its history. As I was finishing this book, I learned that the Waldorf Astoria was to undergo a major renovation by its new owners, Anbang Insurance Company of China, and would be closed for several years while the work was carried out. It suddenly dawned on me that in all the times I've been to the Waldorf, I had never spent a single night there. With unknown changes in the offing, I scrambled to book a room for my husband, Jim, and myself. We reserved a suite in the Waldorf Towers for a single night, Saturday, October 29, to Sunday, October 30, 2016.

The rooms were majestic, with high ceilings, gorgeous furnishings, an incredibly comfortable bed, and a deep, wonderful tub. The views from each window were spectacular. To the south was the sparkling Chrysler Building. To the east was street after street leading to the East River and beyond to Roosevelt Island and the tip of the Four Freedoms Park. To the west was block after block of hotels, office buildings, Broadway theaters, residential towers, and finally the Hudson River. We joined a historical tour of the hotel, led by guide Karen Stockbridge, who told stories of the hotel's origin, the people who built it, who ran it, who lived in it. She spoke of presidents, diplomats, entertainers, and the artisans who created everything that surrounded us.

We ate dinner at the Bull and Bear and the next morning had the Sunday brunch at Peacock Alley while we listened to Emilee Floor play on Cole Porter's piano.

Famous guests at the Waldorf. *Leslie Day*

Peacock Alley Sunday brunch. *Waldorf Astoria Archives*

I couldn't help but wonder how the hotel might be different when I next visited. Where will all the historical artifacts go? Will Cole Porter's piano fill the air with music ever again? What about the thousands of photographs of accomplished and celebrated men and women from the late nineteenth through the twenty-first centuries that line the walls? There are photographs of chefs celebrating the end of Prohibition in 1933 and photographs celebrating the end of World War I.

What will happen to the books, letters, and photographs of every president since Herbert Hoover who stayed at the hotel? Will all of this cultural treasure be saved? Marilyn Monroe had a suite here. Dwight Eisenhower and Douglas MacArthur lived here. Journalists, authors, playwrights. dignitar-

Oscar and his chefs celebrating the end of Prohibition, 1933. *Waldorf Astoria Archives*

General John J. Pershing inspecting his troops from the balcony of the Waldorf-Astoria, 1919. Waldorf Astoria Archives

ies, politicians, and hundreds of thousands of men, women, and children from around the world have spent time in this gorgeous, enormous, sparkling palace. Then I began to consider the cuisine prepared by world-famous chefs with spices, herbs, and fruit from the rooftop kitchen garden and honey from its honeybees. I didn't know whether the new owners would replace the garden. Our window looked down nine stories to the garden below. Only about eight herb beds remained. The outdoor carpet lay discolored where the six hives had stood.

The bees had been moved to Andrew Coté's farm in Connecticut. Next summer they would be moved again to the Hilton Hotel on 6th Avenue and West 53rd Street. Even though the Waldorf Astoria Hotel had new owners, the Hilton Corporation would continue to manage it for the next century, as they had managed it since 1949, when Conrad Hilton acquired management rights, calling the Waldorf "the greatest of them all."

I wrote this book impelled by the thought that we all need to celebrate the generous people at the Waldorf Astoria and St. Bartholomew's Church. In

Remains of chefs' garden. *Leslie Day*

writing it, I learned a lot about the human spirit and the spirit of the hive; the supportive relationships among human neighbors, and bees in the hive, and between bees and flowering plants. I found myself occasionally awestruck by the dedication of the bees to their queen and their siblings. I came to see the hotel like the queen, supported by the thousands of people who care about and for her. The importance of community, the interdependency of living beings within each community had come to life for me.

The head concierge, Michael Romei, has worked at the Waldorf for twenty-three years. He told me that the Waldorf is part of his body and soul, and he is part of it. For decades, he has helped diverse individuals: North American, South American, African, European, Asian, South Asian, Australian, Pacific Islanders, Russian, and Balkan who have stayed or worked there. He is a part of this caring group of people who keep this gorgeous building running. He is part of a continuum that has, for over eighty years, kept the burled wooden paneling of her elevators shined to a high luster, cleaned every mirror every day, made each chandelier sparkle, and served delicious food to millions of people.

What will happen to the art deco metalwork that is considered a tour de force of silver nickel craftsmanship? It is everywhere you look: elevator doors, mailboxes, intricate window grilles, handrails, and the ceiling of the Starlight Roof.

And what will happen to the artwork in the Park Avenue lobby, including the magnificent mosaic by French artist Louis Rigal. Rigal received fame when his work was shown at the 1925 Paris Exposition Internationale des Arts Décoratifs et Industriels Modernes (the International Exhibition of Modern Decorative and Industrial Arts). The term *art deco* was coined from this exposition. Craftsmen from the V. Foscato Company used 148,000 marble tesserae for the *Wheel of Life* mosaic and the floor surrounding it in the Park Avenue lobby. Marble for the floor and the mosaic was imported from around the world: Italy, France, Belgium, Africa, Ireland, Greece, and Turkey. Each piece was cut by hand by these mosaic artists. It took the workers eight years,

Art Deco metalwork, elevator door. *Leslie Day*

1931 through 1939, to assemble this mosaic. During the day, the floor was covered with a rug designed by Rigal, depicting the same *Wheel of Life*, show-

Youth and Friendship, section of Wheel of Life mosaic by artist Louis Rigal. *Leslie Day*

French art deco artist Louis Rigal standing before his illustrated plan for the *Wheel of Life* mosaic, 1935. *Waldorf Astoria Archives*

ing the joy of birth, youth and friendship, life's struggles, maturity, old age, and, finally, death.

And throughout each night for eight years, the rug was removed and the artisans laid down the tesserae until dawn when the rug was laid down again to protect the evolving mosaic masterpiece.

In the 1960s and 1970s, the hotel décor was modernized and the *Wheel of Life* mosaic was covered by layers of carpeting. Jim Blauvelt, the executive director of catering who has been with the hotel since the early 1980s, described the flood that led to the discovery of Rigal's mosaic:

> Around 1983 we were doing a reception for the President of Pakistan in the Vanderbilt Room, and an old pipe in the fire extinguisher system above the ceiling burst. All that rusty water that sat in the system for years, waiting for the sprinklers to go off, poured down into the room. Those sprinklers went off with gusto. The ceiling sagged and parts of the plaster fell. It flooded the Vanderbilt Room, which is a terraced room. Smelly, rusty water cascaded out of the Vanderbilt Room, down the steps, into the Park Avenue lobby and then down the Park Avenue lobby steps right out onto Park Avenue. It was quite a mess.

Blauvelt said that at first they tried to wet vac the carpeting in the lobby, but there was too much water. "And when we started to remove the carpeting we found a carpet underneath and then another carpet and then another carpet and underneath it all we found the Rigal mosaic that had been covered over for many, many, many years."

He recalled that "in the United States in the 1960s and 1970s interior design was at an ebb with shag carpeting and lots of rust and yellow and burnt orange décor," and it was during that time the interior designer at the Waldorf covered over the Park Avenue lobby murals with blue velvet drapes and gold tassels. These thirteen allegorical murals, also by French artist Louis Rigal, depict beautiful muscular men hauling in nets filled with fish and gorgeously gowned women luxuriating in gardens. The interior designer also covered Rigal's exquisite *Wheel of Life* mosaic with carpeting, and all of the art deco nickel-plated grilles on the windows with drapery.

So this leak and the removal of the carpet in the early 1980s spurred interest into not only what was underneath the carpeting, but also what was behind the curtains and the drapery. The Hilton Corporation funded the restoration of the Park Avenue lobby. The carpets were removed revealing Rigal's *Wheel of Life* mosaic. The blue drapes were removed revealing the Rigal murals. The curtains were taken down revealing the nickel-plated grilles. The silver lining was that this flood exposed this unbelievable space. The whole Park Avenue lobby is filled with treasures and that began an eight-year restoration of the hotel in the 1980s and into the early 90s. At the time, this was a 350-million-dollar capital project, which in 1980s dollars was an enormous sum of money. We did the hotel top to bottom. But it was that flood that started the whole thing.

When the Waldorf closes this time, two-thirds of the hotel's almost 1,500 rooms will become condominiums. When word got out that the historic Waldorf was going to be drastically changed, there was a strong desire throughout the city's preservation communities to protect the magnificent public spaces within the hotel. A public outcry to protect art deco treasures in the hotel's public spaces, including Rigal's mosaic and murals, led the New York City Landmarks Preservation Commission to consider landmarking the Waldorf's interiors before construction began.

Included within these public spaces is the Grand Ballroom. When it was first built, this ballroom, at 35,000 square feet and four stories tall, was the largest space of its kind. Its tiers of private boxes are decorated with art deco bas-relief, and its walls are covered with gazelles, fountains, leaves, and birds.

The art deco metal and etched glass grilles conceal organ pipes from a Moller orchestral organ, considered one of the finest in the world. This is where Canadian American bandleader Guy Lombardo and his band, the Royal Canadians, helped Americans ring in New Year's Eve live on television from 1954 to 1976. Louis Armstrong played his last public performance there. Every four years, presidential candidates roast each other at the Catholic Charities of New York Alfred E. Smith Memorial Foundation Dinner.

The paintings, panels, moldings, and frescoed ceiling of the Basildon Room, another public space within the hotel, were originally part of a salon in Basildon Park, an eighteenth-century estate in Berkshire, England, now

Ella Fitzgerald
and the Count
Basie Orchestra,
Starlight Roof,
1956. *Waldorf
Astoria Archives*

Silver Gallery
with Edward
Simmons murals.
*Waldorf Astoria
Archives*

preserved by the UK's National Trust. In 1929, the then owner, a real estate developer, sold off parts of the house to the Waldorf, the Metropolitan Museum of Art, Louisiana State University Museum of Fine Art, and the Boston Museum of Fine Arts.

Other magnificent public places in the Waldorf Astoria considered for preservation include the Starlight Roof, where Ella Fitzgerald, Frank Sinatra, and other enormously talented singers performed. The main lobby where the 1893 World's Fair Clock stands, Peacock Alley, the Jade Room, the Astor Gallery, the Vanderbilt Room, and the Empire Room, where Lena Horne, Count Basie, Gordon MacRae, Victor Borge, and other performers appeared regularly.

The Silver Gallery has murals by Edward Simmons of the seasons and months of the year. These murals were in the Waldorf Hotel on 5th Avenue and are some of the few art pieces that remain from the original hotel.

The main lobby of the Waldorf with its great clock that sat in the lobby of the original hotel is a magnificent space. The ceiling is adorned with art deco medallions, whose designs of animals and foliage are made of pounded gold and silver metal. Directly above the 1893 Goldsmiths and Silversmiths Clock is a medallion with a woman holding a theatrical mask surrounded by golden foliage. Erin Allsop, former archivist at the Waldorf, wrote in the hotel blog that "these medallions blend the Neoclassical elements of Art Deco style with a nod to the theater, an integral part of the culture of New York City."

In July 2016, when it became known that the Waldorf would undergo extensive renovations during a three-year closure, the New York Landmarks Conservancy, the Art Deco Society of New York, and the Historic Districts Council started a campaign to preserve the Waldorf's interiors.

In the article *The Waldorf Astoria Interiors, A Case for Preservation*, Meghan Weatherby, executive director of the Art Deco Society wrote:

> New Yorkers and preservation advocates everywhere were distressed when The Wall Street Journal reported in June 2016 that the hotel's new owner, Beijing-based Anbang Insurance Group Co., planned to "gut the hotel and convert as many as 1,100 rooms into private apartments." The Art Deco Society of New York believes that the Waldorf Astoria must be preserved for both its architectural merit and its importance in the evolution of New York City as a center of world culture. . . . Being deprived of the Waldorf's important public interiors would be a blow to the architectural, political, and social history of New York City and the global community of Art Deco lovers.

The Historic Districts Council sent a letter to the Commission urging

> the Landmarks Preservation Commission to designate the interiors of the Waldorf Astoria at 301 Park Avenue, Manhattan. These interiors were designed to be the utmost in hotel opulence and survive remarkably intact. A major restoration was undertaken

in 1983. The rooms that need protection include the Park Avenue lobby, which includes the "Wheel of Life" mosaic tile artwork by the 1925 Paris Exposition showcase artist Louis Rigal, and is composed of 148,000 marble tiles from seven different countries; the Lexington Avenue Lobby and Peacock Alley, which features a clock from the Columbian Exposition of 1893; the Grand Ballroom, Astor and Jade ballrooms, all of which retain their original finishings; the Basildon Room with finishings imported from an 18th century country house in Berkshire, England; the Silver room, which is covered with mirrors, inspired by Versaille's Galerie des Glaces, include Edward Simmons murals of the months and seasons, originally installed in the Astor Gallery in the original 1897 Waldorf on 5th Avenue; and the Starlight Roof, whose original remaining features are largely intact, including the Art Deco grille work ceiling, which originally retracted, allowing guests to dine and drink beneath the stars above Park Avenue before being permanently closed in 1950 to accommodate HVAC. Original features abound throughout all of the major, publicly accessible rooms and corridors in this magnificent building, including Art Deco moldings, ceiling medallions, elaborate carved woodwork, marble pilasters, murals, grille work, railings, light fixtures, banisters, counter tops, door enframements and plaster work. Don't let the Waldorf Astoria be destroyed on your administration's watch. The Waldorf's interiors are irreplaceable artworks which add to the glory of New York City and should be preserved for future generations of New Yorkers.

I'm one of those people who believe that protection is necessary, and there are many forward thinkers out there who would understand that change doesn't need to mean demise.

When the venerable Waldorf Astoria reopens, I know it won't be exactly the same as what I experienced in October 2016. In 1986, the Hilton Corporation hired Kenneth Hurd & Associates to renovate this revered hotel, and it changed even as its spirit was preserved. In a video made by the Waldorf in 2007, Kenneth Hurd said, "The history of New York is embodied in this hotel. It is a cultural reflection of our history in the 20th century."

I would say that the history of the Waldorf Astoria is a cultural reflection of our history in the late nineteenth through the early twenty-first centuries.

My hope is that the spirit of the Waldorf will be preserved when she reopens the next time and that she will be the thriving hive of culture, beauty, and human history that she has been for 125 years. And maybe she will have a future where even honeybees and their flowers can live and flourish. But, on this winter day in 2016, the honeybees have been removed and the raised flower beds contain just a few remnants of the plants used by the chefs. Very soon the beds will be covered with snow, and spring will bring no harvest.

Maybe it's just fear of the unknown that's gripped me. Maybe I need to focus on all the special times this hotel has seen—on what did exist, even for a brief time: the garden and honeybees and the special relationship with St. Bart's. My mind whirls from Frank Sinatra and Cole Porter, then to Chefs Garcelon and Betz serving dinner at St. Bart's.

In March 2017, I called my editor, who recognized my number and began with his usual greeting, "I hope you're calling to tell me the book is finished?"

"Almost," I responded, "but I've got some good news."

When news of the closing and conversion of many of the rooms to condominiums came in, I had called him as a distraught author. I had envisioned readers going up to see the garden and the bees. That was impossible now. At that time he had said to me, "Well, at least you know how your book ends: sadly."

Maybe not. The New York City Landmarks Preservation Commission ruled in March 2017 that the developer would be required to preserve some of the Waldorf's history. Here's their heartening press release, followed by the response from Anbang:

> The New York City Landmarks Preservation Commission today unanimously voted to landmark several interior spaces of the Waldorf-Astoria Hotel in Manhattan. Considered one of New York City's most prominent and culturally significant hotels, the Waldorf Astoria was designed by Schultze and Weaver and opened in September 1931. The hotel's exterior became a New York City landmark in 1993. Today's vote will protect the building's most lavish public spaces, including interconnected rooms and corridors on the ground, first, second and

Grand Ballroom.
Waldorf Astoria Archives

Art deco metalwork on
elevator doors leading
to the Grand Ballroom.
Waldorf Astoria Archives

third floors. The designation brings the number of New York City Interior Landmarks to 119.

"The Waldorf Astoria Hotel has some of the most internationally renowned rooms in all of New York City," said Commission Chair Meenakshi Srinivasan. "These designated spaces connect seamlessly from the foyers and lobbies on the ground and first floor to the Grand Ballroom and other public rooms on the third floor. Today's action not only protects the rich and beautifully detailed art-deco features of the hotel's interior public spaces, it also preserves the

Basildon Room ceiling fresco. *Leslie Day*

unique experience of moving through the hotel's varied interiors, which countless New Yorkers and visitors have enjoyed for more than eight decades."

Designated spaces on the first floor include the lofty Park Avenue Lobby, an entry hall with thirteen murals and an immense floor mosaic by the French artist Louis Rigal, and the elegant wood-paneled Main Lobby, featuring black marble pillars and ceiling reliefs. On the east end of the first floor, near Lexington Avenue, are elevators with distinctive metal doors and double staircases with "frozen fountain" balustrades leading through the first, second and third floors.

On the third floor, designated spaces include the glittering Silver Gallery, a long mirrored hallway with a black-and-white mosaic floor that links four ballrooms. The Silver Gallery has a coved ceiling that incorporates twelve murals by the American artist Edward Emerson Simmons. The murals, which depict the months of the year, are among the only features salvaged from the hotel's original Fifth Avenue building. The Grand Ballroom, one of the largest event spaces in New York City, can accommodate as many as 1,550 guests. Arranged on three levels, the space has projecting balconies and an elaborate ceiling relief. This fabled room has hosted countless dinners, banquets, galas and balls, including the annual Alfred E. Smith Dinner, a fundraiser for Catholic charities that attracts major presidential candidates.

Also on the third floor is the Basildon Room, which has colorful wall and ceiling panels acquired from an eighteenth-century British mansion, as well as the more restrained Jade Room and Astor Gallery.

Art deco metalwork. *Waldorf Astoria Archives*

Chef David Garcelon talking to senior citizen group visiting the honeybee garden.
Leslie Day

While these memorable interiors have had alterations and some-times exhibit new finishes, most retain their original dimensions and share such unifying elements as gilded plaster reliefs, nickel-bronze metalwork, glazed doors, mirrored walls and various types of exotic wood paneling.

Over the years, the hotel's art deco and Modern Classical–style motifs create strikingly elegant spaces that have played host to an impressive roster of guests, including numerous political leaders and popular entertainers. The Waldorf-Astoria Hotel is one of two major early 20th-century hotels in New York City that have preserved many of their original public spaces. While the Plaza Hotel, which had its interiors designated in 2005, features Renaissance Revival style decoration, the Waldorf is notable and unique for containing some of the finest and most culturally significant Art Deco interiors in New York City.

When the new owner of the hotel, Anbang Insurance, learned of the Landmarks Preservation Commission's vote concerning the hotel's interior spaces, they released this statement: "Anbang knows the Waldorf Astoria's history is a large part of what makes this hotel so unforgettable. That is why we fully supported the commission's recommendations for designation of the Waldorf Astoria's most important public spaces and applaud the commission on achieving landmark status for them."

I found the response encouraging. I look forward with great hope to entering the Waldorf when it reopens in a few years. I am relieved that Fare Share Friday continues at St. Bart's on the Friday after Thanksgiving with the continued support of the major hotels, and that the chefs I am so fond of all found wonderful jobs where they can continue to create their magic.

David Garcelon has opened a new hotel for the Fairmont Group in Austin, Texas. Peter Betz is now the executive chef at the InterContinental New York Barclay, across the street from the Waldorf. A magnificent hotel built around the same time as the Waldorf, the Barclay also has a rooftop apiary. Charlie Romano is executive pastry chef for the Peninsula Hotel in New York City.

In finishing this book I have learned that the Chinese government has taken over the Anbang properties and is now in control of the Waldorf. I am hopeful that when I next walk through the doors of the hotel I will feel enormous relief and gratitude to find the preserved public spaces. I also look forward to seeing the changes that are made. But mostly what I expect to feel is that I'm still in a special place, where once upon a time David, Peter, Charlie, the staffs of the Waldorf Astoria and St. Bartholomew's Church united with a garden filled with honeybees to show the world the real New York: a city of creative and compassionate people where nature can thrive. That's the true heart of my beloved city.

WALDORF ASTORIA KITCHEN AND PEACOCK ALLEY BAR RECIPES USING HONEY

The recipes that follow were developed by chefs discussed in this book. Each of them has honey as an ingredient. From 2012 to 2016, the honey used in these recipes came from the hotel's rooftop hives.

PREVIOUS PAGE: Top of the Waldorf Honey jar. *Leslie Day*

WALDORF SALAD
(OSCAR TSCHIRKY, MAÎTRE D'HÔTEL)

The Waldorf Astoria's most famous dish, this salad was created in 1893
when the original Waldorf Hotel opened its doors. The salad is
deliciously crunchy, with celery, fruit, and nuts.

INGREDIENTS

⅓ cup mayonnaise
¼ cup sour cream
1 tablespoon fresh lemon juice
1 teaspoon honey
¼ teaspoon salt
3 red apples
2 stalks celery
½ cup walnuts
⅓ cup dark seedless raisins (optional)

In a medium bowl, with wire whisk, mix mayonnaise, sour cream,
lemon juice, honey, and salt until blended. Add apples, celery,
walnuts, and raisins, if using the latter, to dressing in the bowl;
toss until mixed and evenly coated with dressing.

HONEY VINAIGRETTE
(DAVID GARCELON, DIRECTOR OF CULINARY)

Here is a simple, yet delicious, salad dressing.

INGREDIENTS

1 cup oil
1 cup olive oil
⅔ cup vinegar
¼ cup honey

Place honey and vinegar in a blender. Slowly add oils until mixed.

ROYAL YORK HONEY GLAZED FIGS AND GOAT CHEESE
(DAVID GARCELON, DIRECTOR OF CULINARY)

INGREDIENTS

2 fresh figs cut into ½-inch rounds
2 ounces goat cheese, cut in slices
(ideally, Mariposa Ontario goat cheese)
1 tablespoon butter
1 tablespoon honey
4 slices baguette, toasted
1 strip maple smoked bacon (cooked crisp)

Place figs on a baking tray, each with 1 slice of goat's cheese, butter,
a drizzle of honey, and ¼ of the crisp bacon, warm through for
3–4 minutes in a 350°F oven. Place baked figs on baguette and drizzle
with remaining honey. Serve immediately. Makes 4 servings.

HONEY BRIOCHE BREAD
(DAVID GARCELON, DIRECTOR OF CULINARY)

Honey can add a special touch to the taste of bread. This recipe will provide you with a fresh loaf that has a subtle honey flavor.

INGREDIENTS

½ cup milk
¼ cup honey
¾ ounce fresh yeast
7 tablespoons butter (soft)
4 eggs
½ teaspoon salt
4 cups flour

Heat milk and honey until lukewarm, then add yeast and allow to sit in a warm place. Beat eggs and butter together, then stir into milk mixture. Mix in the flour and form dough, knead until it becomes shiny and smooth. Refrigerate dough for 30 minutes, then leave in a warm place until the dough doubles in size. Knead dough again. Shape into a loaf and place in a bread loaf pan. Return the dough to a warm place and allow to rise again. Bake for 45 minutes at 425°F.

HONEY LAVENDER CHICKEN JUS
(PETER BETZ, EXECUTIVE CHEF)

Using honey from the rooftop bees and lavender and thyme from the
rooftop garden, the chefs created a mouthwatering glaze.

INGREDIENTS

2 chicken thighs (cut into pieces)
2 chicken legs (cut into pieces)
1 cup onion sliced
1 cup carrot chopped
½ cup celery chopped
3 sprigs fresh thyme
2 tablespoon local honey
1 teaspoon fresh lavender (picked)
1 bulb of garlic (cleaned)
6 ounces white wine
3 cups chicken stock
Blended oil
1½ tablespoons room temperature butter

Caramelize the chicken, carrots, celery, and garlic in a pot until brown.
Add the wine and deglaze the pan until the bottom is clean. Reduce
the wine until almost completely evaporated and add the chicken stock.
Add the thyme, lavender, and honey. Cook on low flame until reduced
by half. Remove and strain through a fine china cap strainer or chinois
strainer. Return the liquid to a pot, and on a very low flame, add the
butter slowly until thick. Season with salt and pepper.

"TOP OF THE WALDORF" ROOFTOP HONEY GLAZED CHICKEN SATAY

(DAVID GARCELON, DIRECTOR OF CULINARY)

INGREDIENTS

6 chicken breasts

1 bulb fresh whole garlic, smashed

10 sprigs fresh thyme

4 large shallots, sliced

2 lemons sliced in rounds

1½ cup extra virgin olive oil

3 cups chicken stock

1 cup honey

24 wooden skewers soaked in water for 30 minutes
to prevent scorching

Split chicken breasts in half horizontally and then cut into 2 ounce strips
(approximately 24 pieces). Mix olive oil, lemons, shallots, thyme, and
garlic. Marinate chicken with mixture for 4–6 hours (24 hours for best
results). Remove chicken pieces from marinade and skewer each with a
soaked skewer. Place in refrigerator until ready to grill. Add stock to saucepan and reduce on medium heat until stock coats back of a spoon. Remove
from heat. Whisk honey into reduced stock and season with salt and pepper
as desired. Reserve half for glaze, half for dipping sauce. Season chicken
with salt and pepper and grill satays on barbecue grill until nearly done.
Glaze satays with honey glaze and return to warm grill for 30 seconds until
glaze becomes slightly caramelized. Serve satays warm with reserved honey
glaze as dipping sauce.

LAVENDER HONEY ICE CREAM
(CHARLIE ROMANO, EXECUTIVE PASTRY CHEF)

The lavender adds a light flavoring that mixes well with the sweetness of the honey. It makes a great topping over berries, waffles, or shortcake.

INGREDIENTS

1 quart heavy cream

2 ounces sugar

8 ounces honey

12 egg yolks

½ ounce fresh, dried lavender

Bring honey to a boil and gradually add milk. Add lavender florets to the honey-milk mixture, and then reduce the mixture to a simmer. Because lavender has a very strong essence, you should avoid using more than the recommended amount. Separately combine the sugar with the egg yolks. Add a touch of the simmering honey-milk mixture to prevent the sugar from coagulating the egg yolks. Whisk the sugar and egg yolk mixture very well. Once the honey-milk mixture begins to simmer, gradually add a third of it into the sugar and egg yolk mixture, making sure you whisk it thoroughly. This will temper the sugar and egg yolk mixture before it is combined with the hot milk mixture. Once fully tempered, add the remaining sugar and egg yolk mixture into the hot honey-milk mixture. Lower the heat to medium, and continue to cook this mixture, making sure you stir it thoroughly with a spatula or a wooden spoon. As it coagulates, continue to cook it to about 180°F (82°C), or *nappé* stage (test by dipping in a metal spoon and seeing whether it stays coated after you lift it away from the mixture). When the *nappé* stage is reached, strain it through a *chinois* (a metal conical sleeve with extremely fine mesh) and place the hot mixture in a pan in an ice bath to stop the cooking.

When cool, place the mixture in the refrigerator and allow to mature overnight. The next day, fully stir the mixture to make it homogeneous, and spin in an ice cream machine, or use an ice cream machine substitution method that can be found on the internet. Do not overspin because of the high fat content in heavy cream.

HONEY TRUFFLE (GANACHE)
(CHARLIE ROMANO, EXECUTIVE PASTRY CHEF)

INGREDIENTS

10 ounces heavy cream
6 ounces honey
1 vanilla bean, scraped
32 ounces dark chocolate, chopped

Combine heavy cream, honey, and scraped vanilla bean in a saucepot and bring to a rolling boil on a stove top. Stir occasionally to ensure it won't scorch on the bottom. Place the chopped chocolate into a heat-resistant bowl and set aside. When the cream mixture is boiling, take it off the heat and pour all of the hot cream mixture onto the chopped chocolate pieces. Let this mixture rest for 1 minute. Once the chocolate becomes molten, stir it with a rubber spatula or wooden spoon from the inside out to create an emulsion. When the center begins to come together, gently stir toward the outside to incorporate the remaining liquid heavy cream. After everything is incorporated, use a hand blender, if necessary, to homogenize. Place plastic film wrap directly onto the surface of the ganache to ensure that the surface is sealed. Keep at room temperature until set and ready to be served. Keep in the refrigerator tightly sealed if it is to be kept for up to 1 week.

HONEY TEA LOLLIPOPS
(CHARLIE ROMANO, EXECUTIVE PASTRY CHEF)

INGREDIENTS

1 cup sugar

6 tablespoons honey

2 tablespoons glucose

8 tablespoons brewed tea

Nonstick cooking spray

Lollipop molds

Lemon rinds

Mint leaves

Lollipop sticks

Combine the sugar, honey, glucose, and brewed tea in a small saucepan fitted with a candy thermometer and cook on high heat. While cooking the candy syrup, occasionally wash down the pan sides using a clean brush dipped in water to prevent crystallization. Without stirring, cook until mixture reaches 310°F degrees, or the hard crack stage, on the candy thermometer. Spray lollipop molds lightly with nonstick cooking spray. Place lemon rinds and mint leaves in each section. Place lollipop sticks into mold. Remove the pot from the heat and dip it into an ice bath for 15 seconds to stop the cooking. Remove the pot from the ice bath. To avoid air bubbles, swirl hot candy syrup in both directions, being careful not to overmix. Pour the syrup into the molds two-thirds of the way full and cool at least 20 minutes. Remove from molds.

MULTIGRAIN BAR
(CHARLIE ROMANO, EXECUTIVE PASTRY CHEF)

Here's the perfect snack, one you can carry along to be eaten on the run, or pass around the table after a casual meal.

INGREDIENTS

5¾ ounces Kellogg's Special K cereal
6¾ ounces old-fashioned oats
7⅞ ounces steel-cut oats
2½ ounces pumpkin seeds
2½ ounces sunflower seeds
3½ ounces dried cherries
5½ fluid ounces honey
1½ fluid ounces corn syrup
5½ fluid ounces orange juice
2⅛ ounces apricot jam
¼ teaspoon vanilla bean paste
Zest of ¼ orange
1 ounce butter

Combine dry ingredients and set aside. Combine wet ingredients to form a sugar mixture: honey, corn syrup, orange juice, apricot jam, and vanilla bean paste and cook to 235°F (113°C). Add butter and cook to a full boil. Add dry ingredients to the boiling liquid and fully coat with sugar mixture. Pour onto greased sheet pans with low frame. Let cool and cut to 1½ inches by 2½ inches.

HONEY NOUGAT GELATO
(CHARLIE ROMANO, EXECUTIVE PASTRY CHEF)

This is one of those rich-tasting desserts that truly melts in your mouth. The addition of the honey almond brittle adds both sweetness and crunch.

INGREDIENTS

For the gelato:
1 cup heavy cream
3 cups whole milk
2 ounces sugar
7⅛ ounces honey
12 egg yolks
For the honey almond brittle:
4 ounces roasted sliced almonds
2 ounces honey cooked to dark amber color

Prepare the gelato:

Bring the honey to a boil and gradually add the milk and heavy cream. Bring this milk mixture to a simmer. Combine the granulated sugar with the egg yolks. Add a touch of the milk mixture to prevent the sugar from coagulating the egg yolks. Whisk the egg yolk mixture very well. Once the milk mixture begins to simmer, gradually add one-third of this liquid into the egg yolk mixture, making sure to whisk it thoroughly. This will temper the egg yolk mixture before it is combined with the hot milk mixture. Once fully tempered add the liquid egg yolk mixture into the hot milk mixture. Lower the heat to medium and continue to cook this mixture, making sure to stir it thoroughly with a spatula or a wooden spoon. As it coagulates, continue to cook it to 180°F (82°C), or *nappé stage* (test by dipping in a metal spoon and seeing whether it stays coated after you lift it away from the mixture). When the desired stage is reached, strain it through a *chinois* (a metal conical sleeve with extremely fine mesh): and place the hot mixture in an ice bath to stop the cooking. When cool, place the mixture in the refrigerator and allow it to mature overnight. The next day, fully stir the mixture to make it homogeneous and spin in an ice cream machine or use a similar method to achieve the result. Do not overspin because of the high fat content in heavy cream.

Prepare the almond brittle:
Cook honey until an amber color is achieved and fold in the sliced almonds. Once the almonds are fully coated, pour the honey mixture onto a baking sheet and allow to cool. After it has cooled, break it into pieces and fold into the ice cream in a large bowl.
Note: the honey brittle can also be used as topping on a quality, store-bought ice cream.

HONEY MADELEINE
(CHARLIE ROMANO, EXECUTIVE PASTRY CHEF)

INGREDIENTS

31¾ ounces sugar

Zest from 7 lemons

15 eggs

6⅝ fluid ounces milk

33 ounces pastry flour

1 ounce sifted baking powder

1¼ fluid ounces lemon juice

13¾ fluid ounces melted butter

13¼ fluid ounces honey

Whip the sugar, lemon zest, and eggs to ribbon stage. Add everything else in the order given in the recipe. Allow the batter to rest in the refrigerator for at least 3 hours. Lightly butter a metal madeleine mold. After the 3 hours of cooling, use a pastry bag to pipe the batter into the molds halfway up to the rim. Bake for 12–14 minutes at 340°F (170°C) in a convection oven, or at 375°F (190°C) in a nononvection oven. Prepare the recipe one day in advance of serving.

WALDORF ASTORIA COCKTAILS

Oscar Tschirky putting liquor bottles in wood chipper at the start of Prohibition, 1920.
Waldorf Astoria Archives

In 1931, as the dry days of Prohibition dragged on, the old Waldorf closed, the new hotel rose on Park Avenue, and newspaperman and cultural historian Albert Stevens Crockett published the *Old Waldorf Bar Days* in tribute to the "big old Brass Rail bar" in the original hotel. In 1934, after the repeal of Prohibition, he wrote *The Old Waldorf Astoria Bar Book*. More than eighty years later, Waldorf Astoria Peacock Alley beverage manager, mixologist, and cocktail historian Frank Caiafa published *The Waldorf Astoria Bar Book* (Penguin, 2016) filled with interesting Waldorf stories and recipes. The remainder of this section is devoted to cocktail recipes made with the hotel's rooftop honey.

HONEY SYRUP

A standard recipe used by bartenders to add a unique sweetness, aroma, and flavors of flower nectar to cocktails.

INGREDIENTS

4 ounces honey
4 ounces hot water

Add ingredients to saucepan and stir over medium heat until honey is dissolved. Remove from heat, let cool, and store in clean glass bottle. Refrigerate up to 2 months.

BEES KNEES

A classic dating back to Prohibition, when the aroma of honey and lemon juice helped mask the smell of alcohol and when the phrase "bees knees" meant "the best!" Caiafa joked, "This recipe borrows the popular phrase from the time to label a cocktail that generates a little buzz."

INGREDIENTS

2 ounces Citadelle gin
¾ ounce house-made honey syrup (see recipe on p. 164)
¾ ounce fresh lemon juice
1 ounce butter

Add all ingredients to mixing glass. Add ice and shake well. Strain into chilled cocktail glass. Garnish with lemon twist.

LEAVES OF GRASS

This drink's title is an homage to the poet Walt Whitman and his most-celebrated collection of poetry, *Leaves of Grass*. At the Waldorf Astoria, it is garnished with a single blade of bison grass.

INGREDIENTS

2 ounces Zubrowka Bison Grass Vodka
¾ ounce Quinta do Noval 10 Year Old Tawny Port
½ ounce honey syrup (see recipe on p. 164)
¼ ounce fresh lemon juice

Add ingredients to mixing glass. Add ice and shake well.
Strain into chilled cocktail glass.

THE BUMBLED BEE

In 2015, at the Battle of the Bees, an annual charity event held at the Waldorf Astoria where honey from hives on New York City hotels, stores, museums, and schools competed for best honey, this cocktail represented the Waldorf Astoria. Caiafa told me that he "used local rye whiskey to keep with the theme. To my palate, rye makes for a more balanced result (even an entry-level Armagnac does the trick) though if you use bourbon, it may be a more winning crowd pleaser. Do try both."

INGREDIENTS

1½ ounce New York Distilling Co.'s Ragtime Rye Whiskey
¾ ounce Salers Gentiane Apéritif
½ ounce Noilly Prat Extra Dry Vermouth
¼ ounce honey syrup (see recipe on p. 164)
2 dashes Angostura Bitters
1 dash Regans' Orange Bitters No. 6

Add all ingredients to mixing glass. Add ice and stir for 30 seconds. Strain into chilled cocktail glass. Lemon peel, for garnish.

MARTELL COCKTAIL

Named for what we can only assume to be Oscar's favorite cognac. The recipe is found in Oscar Tschirky's *100 Famous Cocktails* (1934). Caiafa suggests that "although a particular brandy is the namesake, you may want to try it with a lighter-bodied rendition to accentuate the honey and allow it to shine through a bit more."

INGREDIENTS

2 ounces Martell VSOP cognac
½ ounce fresh lime juice
½ ounce honey syrup (see recipe on p. 164)

Add all ingredients to mixing glass. Add ice and shake well. Strain into chilled cocktail glass. Garnish with orange peel.

ROOFTOP MAGIC

This unique cocktail was created for a media event at the Waldorf's in-house weekly magic show. Caiafa explains, "The magic part of the cocktail is that we made the bubbles in the beer disappear. . . . Ha!"

INGREDIENTS

2 ounces black cherry–infused bourbon
(1 cup fresh black cherries, split; 1 750-milliliter-bottle
Henry McKenna, bottled in bond bourbon)

1 ounce Cinzano Rosso sweet vermouth

¾ ounce Empire Brewing Company Downtown Brown Ale
or other brown ale (flat)

¼ ounce honey

1 dash Angostura bitters

Place all ingredients in airtight glass container for 7 days. Fine-strain and funnel back into bottle. Best if used within 2 months. Simply pour the beer into a container, and leave it covered in the refrigerator for 2 days. Then use as you would any other ingredient. Add all ingredients to mixing glass. Add ice and stir for 30 seconds. Strain into chilled cocktail glass. Garnish with brandied cherry. Caiafa mentioned that "the original beer was our Waldorf Buzz brown ale, made by Empire Brewing Company, in Syracuse, New York. It incorporated our 'Top of the Waldorf' honey in the fermentation process and was a huge success. The honey syrup was also made with our rooftop honey, culled from our rooftop hives, and the bitters were house-made as well. Despite the use of honey in this cocktail, it is not as sweet as you might think. The assertive, dry brown ale does its job (along with the bitters) of drying out the finish."

ROOFTOP SUNSET CHAMPAGNE COCKTAIL

This easily made yet complex cocktail is best served outdoors, during long summer sunsets.

INGREDIENTS

5 ounces Sparkling Pointe Méthode Champenoise (North Fork)
1 ounce honey syrup (chilled) (see recipe on p. 164)
2 dashes Peychaud's Bitters

Add sparkling wine to flute or coupe. Add honey syrup to mixing glass. Add ice and stir briefly to integrate and chill. Slowly strain into prepared champagne flute or coupe. Carefully drop in the bitters. A pinkish cloud should form.

LOCOMOTIVE (PITCHER/WARM)

Named for the mode of transportation invented in the nineteenth century. This recipe dates back to 1862, from Jerry Thomas' *Bartenders Guide*.

INGREDIENTS

12 cloves

2 cinnamon sticks

1 750-milliliter-bottle red wine

4 egg yolks

4 ounces honey syrup (see recipe on p. 164)

4 ounces Grand Marnier

Add spices and wine to saucepan over medium heat. Simmer for a few minutes (do not boil). Remove from heat and let rest. Add egg yolks and half of the honey syrup to mixing glass; dry-shake for at least 5 seconds. Add remaining honey syrup and Grand Marnier to mixing glass and roll between mixing glass and shaker a few times to integrate. Add to serving pitcher. Fine-strain wine into pitcher and stir to integrate. Serve in ceramic cups. Garnish with freshly grated nutmeg.

ACKNOWLEDGMENTS

When I started this wonderful project, I knew little about honeybees or the marvelous history of the Waldorf Astoria Hotel. First and foremost, I must thank my nephew, Peter Betz, executive chef of the hotel who first told me about the adventure the Waldorf was embarking upon: establishing an apiary and chefs' garden on the twentieth-floor roof, next door to what had been, once upon a time, Marilyn Monroe's suite.

Next in line I shake the hand and give a grateful hug to Chef David Garcelon, the true father of the apiary and chefs' garden, who, having arrived from Toronto's classy Royal York Hotel to the Waldorf, brought his brainchild to build honeybee hives and a thriving kitchen garden to the hotel and, as a result, helped bring us all together.

I am in awe of the exquisite and delicious food the hotel's chefs prepared. Memories of Chef Calogero (Charlie) Romano preparing and serving honey lavender ice cream decorated with sprigs of lavender flowers makes my mouth water to this day. The smell alone was to die for.

Unfortunately, I did not get to taste all of the cocktails sweetened by rooftop honey that Waldorf Astoria's Peacock Alley beverage manager, mixologist, and cocktail historian Frank Caiafa helped me with for this book. I aim to try to work my way through them at home over the coming years.

Andrew Coté, New York City beekeeper extraordinaire, helped me tremendously by letting me shadow him as he moved from frame to frame through the Waldorf's hives, talking about the life of the hives and what he saw in the combs. He was always accessible by phone, email, or text to answer my many questions on honeybees.

The Waldorf Astoria's archivist, Erin Allsop, supported me tremendously in my research on the hotel and was so generous with her time and her knowledge. It was Erin who made the online archive of the hotel's historic photographs and articles accessible to the public on the website http://www .hosttotheworld.com. In the years she was the archivist, she was an absolute treasure. Although the website containing all of the archival photographs has been taken down, I am hoping that it will be restored at some point. Thousands of digitized photographs portraying New York City and world history through the lens of the Waldorf Astoria from the nineteenth century onward should continue to be made available to the public.

Michael Romei was head concierge of the Waldorf Astoria and its historian for decades, and I can't thank him enough for meeting with me, talking about the history of the hotel, and reading my book for accuracy. He is an accessible, generous, and intelligent man.

Juan Guo, director of general management of the US division of Anbang Insurance, helped me acquire permission to use the archived historic photos of the Waldorf Astoria Hotel, and Sherry Yuan Qu, deputy director of the China General Chamber of Commerce, who I worked alongside serving meals on Fare Share Friday at St. Bartholomew's Church, introduced me to Juan. Thank you, ladies, for your important contribution to this book.

I toured the hotel several times with the Waldorf's tour guide, Karen Stockbridge, and learned so much from her about the hotel's history, starting in the 1800s.

I am grateful to Jonathan Stas, the former marketing communications manager at the Waldorf, for giving me permission to use several of his photographs for this book, including the design he created for the Top of the Waldorf Honey jars. And I am, as always, in debt to my friend, the marvelous photographer Beth Bergman, for photographing the honey jars for this book.

The Reverend Edward Sunderand, LCSW, director of St. Bartholomew's Crossroads Community, helped me further understand the relationship between the Waldorf Astoria Hotel and St. Bart's: the supportive role the hotel and the chefs played in helping to feed the people who needed shelter and food, recruit chefs from other hotels, and raise money to support Crossroads' programs at St. Bart's. He also enlisted the wonderful photographer Jeremy Daniel to take photos of the church's guests eating breakfast at its soup kitchen one cold January morning.

Thank you to Carolyn Gargano, creative art director of Saatchi and Saatchi Wellness, who created the print advertisement for Fare Share Friday and reviewed my chapter on the Fare Share Friday event at St. Bartholomew's Church, where the chefs from the Waldorf and several other major hotels prepare a post-Thanksgiving meal for homeless people in the nave of the church. Carolyn is also a board member of Crossroads Community. I thank you, Carolyn, for your creativity and generosity in the volunteer work you do with those who find shelter and food at St. Bart's.

I learned about the history of St. Bartholomew's Church from my enlightening tour with Becca Earley, author of *Holy Light*, the story of the church's stained glass windows. Becca is a tireless advocate for St. Bart's, its magnificent artistic and architectural history, and the wonderful work they do in the community.

I met Katherine Jollon at an Art Deco Society of NY talk on the gorgeous artwork of Hildreth Meière at Temple Emanu-El on 5th Avenue one cold winter night. Katherine's research on Louis Rigal's *Wheel of Life* mosaic has been tremendously helpful.

I am eternally grateful to Thomas D. Seeley, Horace White Professor in Biology in the Department of Neurobiology and Behavior at Cornell University, and author of many research projects and wonderful books on *Apis mellifera*, including *Honeybee Democracy*, who read my manuscript and set me straight on understanding the lives of bees.

And thanks to the delightful entomologist Justin O. Schmidt, author of *The Sting of the Wild*, for reading my bee chapters for accuracy and whose Pain Scale for Stinging Insects helped me understand the evolution of the honeybee's sting and the pain it inflicts.

Thank you to Mai Reitmeyer and Thomas Baione, librarians at the American Museum of Natural History Library, and Madeline Byrne, Vanessa Sellers, and Stephen Sinon at the New York Botanical Garden Library for their kind help.

Dr. Jerome Rozen, curator at the division of Invertebrate Zoology and a professor at the Richard Gilder Graduate School at the American Museum of Natural History, assisted me as we looked through his microscope at the anatomy of an *Apis mellifera* worker bee and compared it to my renderings to make sure that I was accurate.

I have been indeed fortunate to spend time with the tiniest subjects of my books: the honeybees, through the kindness of Louise Desjardins, my friend and sister volunteer gardener at the Heather Garden in Fort Tryon Park. Right after I completed this book Louise set up a top-bar hive on her rooftop, and I asked whether I could help. It has been one of the great delights of my life to so closely observe these miraculous animals and to get to taste the honey they produce. It is so warm—kissed by the sun—and so sweet. Watching them do so many of the tasks I have researched and written about is utterly fascinating. It's like watching a newborn child—every part of their bodies is interesting to me now. I'm enthralled by just watching their little pink tongues sop up any honey spilled on the hive. I am in love!

Louise of the rooftop bees is also a retired book editor, and I asked her to read my book for accuracy. She helped spot some misconceptions I had missed. She also said it was about everything she loves: New York City history, bees, gardens, and food.

A huge thank you to journalist, historian, and the brilliant author of New York City historical novels, my friend Kevin Baker, for reading the book and giving me such sage advice.

Channing Redford, architect, scientific illustrator, and classmate at the New York Botanical Garden's Botanical Illustration certificate program, read the manuscript and made wonderful and helpful suggestions. She also illustrated the gorgeous cover.

To Suzanne Siegel Slater, author of *Tiny New York*, thank you for helping me discover my title, *Honeybee Hotel*.

Several of my friends read the book and made wonderful suggestions. Thank you Trudy Smoke, Lynn McMahill, Lisa Nicolau, and Sammie Smith.

Thank you Gabe Kirchheimer for coming to the Waldorf with me to photograph the bees and the special events. I appreciate your giving me your time and talent.

My dearest friend, Trudy Smoke, artist and linguistics professor at Hunter College, helped me immeasurably by reading my chapters and edits every couple of weeks for a year. She is a true, patient, and loving friend to whom I am forever grateful.

There are not many people about whom you can honestly say changed your life. Vincent J. Burke, my editor at Johns Hopkins University Press for the past fifteen years, is one. Though Vince is retired from Johns Hopkins University Press now, when I first approached him with the idea for this book he mentioned that it was a different kind of writing than I had engaged in before. During the course of our back and forth writing and editing, he pushed me to be less pedantic, and, if nothing else, this adventure has certainly helped me learn to see and to express the big picture of how tiny insects and larger human animals can work with their own species and with each other to make a difference in this world. Thank you, Vince.

The wonderful Tiffany Gasbarrini, my new insightful and very wise editor at Johns Hopkins University Press, has made important contributions to this book, and her incredibly able editorial assistant, Lauren Straley, helped me immeasurably.

Thank you to my longtime copy editor at Johns Hopkins, Andre Barnett, and to Juliana McCarthy, the managing editor, for your input and help with the manuscript.

And finally to my husband, Jim Nishiura, thanks again for not bugging me about the stacks of books on the floor of our apartment, for month after month, year after year. I'm pretty sure our romantic weekend at the Waldorf made up for all of that.

BIBLIOGRAPHY

INTERVIEWS

Erin Allsop, Archivist, Waldorf Astoria Hotel, July 2014

Peter Betz, Executive Chef, Waldorf Astoria Hotel, September 2012, July 2014, April 2016, June 2016

Jim Blauvelt, Director of Catering, Waldorf Astoria Hotel, December 2016

Andrew Coté, Director, New York City Beekeepers Association, July 2014, October 2014, January 2016, December 2016

Laurel Dutcher, Board Member, Crossroads Community Service, St. Bartholomew's Church (*Source:* Steve Sakson, "A Radical Thanksgiving Dinner," *The Episcopal New Yorker*, Fall 2015.)

Becca Earley, Historian, St. Bartholomew's Church, December 2016

David Garcelon, Director of Food and Beverages, Waldorf Astoria Hotel, July 2014, January 2015, September 2016

Carolyn Gargano, Art Director, Saatchi and Saatchi Wellness, November 2014, March 2017

Katherine Jollon, February 2017

Hilda Krus, Director of the GreenHouse Program, Rikers Island Correctional Facility, July 2015

George Pisegna, Director of Horticulture, New York Horticultural Society, April 14, 2015

Calogero Romano, Executive Pastry Chef, Waldorf Astoria Hotel, March 2017

Michael Romei, Head Concierge, Waldorf Astoria Hotel, December 2016

SOURCES

Alston, Frank. *Skeps: Their History, Making and Use.* West Yorkshire, UK: Northern Bee Books, 1987.

Bishop, Holly. *Robbing the Bees: A Biography of Honey—the Sweet Liquid Gold That Seduced the World.* New York: Free Press, 2005.

Brawer, Catherine Coleman, and Kathleen Murphy Skolnik. *The Art Deco Murals of Hildreth Meière.* New York: Andrea Monfried Editions, 2014.

Buchmann, Stephen. *Honey Bees: Letters from the Hive.* New York: Delacorte Press, 2010.

Caiafa, Frank. *The Waldorf Astoria Bar Book.* New York, Penguin Books, 2016.

Casteel, D. B. *The Behavior of the Honey Bee in Pollen Collection.* Washington, DC: Department of Agriculture, Bureau of Entomology, 1912.

Crowninshield, Frank, ed. *The Unofficial Palace of New York: A Tribute to the Waldorf-Astoria.* New York: Privately printed, 1939.

Doherty, John. *The Waldorf-Astoria Cookbook*, with John Harrisson. New York: Bullfinch Press, 2006.

Ellis, Hattie. *Sweetness and Light: The Mysterious History of the Honeybee.* New York: Crown, 2005.

Farrell, Frank. *The Greatest of Them All.* New York: K. S. Giniger, 1982.

Fisher, Rose-Lynn. *Bee.* New York: Princeton Architectural Press, 2010.

Gayle, Margot, David W. Look, and John G. Waite. *Metals in America's Historic Buildings.* Washington, DC: Department of the Interior, National Park Service, 1992.

Gibran, Kahlil. *The Prophet.* New York: Alfred A. Knopf, 1960.

Goodman, Lesley. *Form and Function in the Honey Bee.* Cardiff, UK: International Bee Research Association, 2003.

Grout, Roy, A., ed. *The Hive and the Honey Bee.* Hannibal, MO: Dadant & Sons, 1946.

Heinrich, Bernd. *Bumblebee Economics.* Cambridge, MA: Harvard University Press, 2004.

Horn, Tammy. *Bees in America: How the Honey Bee Shaped a Nation.* Lexington: University Press of Kentucky, 2005.

Hubbell, Sue. *A Book of Bees.* New York: Mariner Books, 1998.

Huber, Raymond. *Honey Bees: An Essential Guide.* Self-published, Amazon Digital Services, 2013.

Jiler, James. *Doing Time in the Garden.* Oakland, CA: New Village Press, 2006.

Jones, Richard A., and Sharon Sweeney-Lynch. *The Beekeeper's Bible: Bees, Honey, Recipes and Other Home Uses.* New York: Stewart Tabori & Chang, Harry N. Abrams, 2011.

Kahr, Joan. *Edgar Brandt: Art Deco Ironwork.* Atglen, PA: Schiffer, 2010.

Kaplan, Justin. *When the Astors Owned New York: Blue Bloods and Grand Hotels in a Gilded Age.* New York: Penguin Books, 2007.

Kritsky, Gene. *The Quest for the Perfect Hive.* New York: Oxford University Press, 2010.

Langstroth, L. L. *Langstroth's Hive and the Honey-Bee: The Classic Beekeeper's Manual.* Mineola, NY: Dover, 2004.

Maeterlinck, Maurice. *The Life of the Bee.* Overland Park, KS: Digireads, 2004.

McCarthy, James Remington. *Peacock Alley.* New York: Harper & Brothers, 1931.

Moisset, Beatriz, and Stephen Buchmann. *Bee Basics: An Introduction to Our Native Bees.* Washington, DC: USDA Forest Service and Pollinator Partnership Publication, 2011.

Morehouse, Ward, III. *The Waldorf-Astoria: America's Gilded Dream.* New York: M. Evans, 1991.

Morison, Guy D. "Memoirs: The Muscles of the Adult Honey-Bee (*Apis mellifer* L.)." *Journal of Cell Science* 1927:511-26.

Nordhaus, Hannah. *The Beekeeper's Lament: How One Man and Half a Billion Honey Bees Help Feed America*. New York: Harper Perennial, 2011.

O'Toole, Christopher, and Anthony Raw. *Bees of the World*. London: Blandford, 1991.

Page, Robert E., Jr. *The Spirit of the Hive: The Mechanisms of Social Evolution*. Cambridge, MA: Harvard University Press, 2013.

Preston, Percy, Jr. "The Portal of Saint Bartholomew's Church in New York City." *Magazine Antiques*, December 2001.

Proctor, Michael, Peter Yeo, and Andrew Lack. *The Natural History of Pollination*. Portland, OR: Timber Press, 1996.

Ransome, Hilda M. *The Sacred Bee: In Ancient Times and Folklore*. London: George Allen & Unwin, 1937.

Schmidt, Justin O. *The Sting of the Wild*. Baltimore: Johns Hopkins University Press, 2016.

Seeley, Thomas D. *Honeybee Democracy*. Princeton, NJ: Princeton University Press, 2010.

Smith, Christine. *St. Bartholomew's Church in the City of New York*. New York: Oxford University Press, 1988.

Snodgrass, Robert. *The Anatomy of the Honey Bee*. Washington, DC: Government Printing Office, 1910.

Stell, Ian. *Understanding Bee Anatomy: A Full Colour Guide*. Middlesex, UK: Catford Press, 2012.

Weatherby, Meghan. "The Waldorf Astoria Interiors: A Case for Preservation." *Journal of the Art Deco Society of New York*, 2016.

Wilson, Joseph S., and Olivia Messinger Carril. *The Bees in Your Backyard: A Guide to North America's Bees*. Princeton, NJ: Princeton University Press, 2016.

Wilson-Rich, Noah. *The Bee: A Natural History*. Princeton, NJ: Princeton University Press, 2014.

Winston, Mark L. *Bee Time: Lessons from the Hive*. Cambridge, MA: Harvard University Press, 2014.

Winston, Mark L. *Biology of the Honey Bee*. Cambridge, MA: Harvard University Press, 1987.

INDEX

Illustrations in the gallery, which follow page 72, are cited using the letter *G*.